From Copernicus to Einstein

This is a lively and lucid explanation of what the world's scientists have discovered about the physical world—down through the ages.

With wit and brilliance, Bertrand Russell penetrates the complexities of mathematics and physics to present the theory of relativity in clear, concise language that can be readily understood by the reader with little scientific training.

Using analogies drawn from everyday experience, he explains the theories of Copernicus, the ether experiments of Michelson and Morley, the Fitzgerald contraction hypotheses as developed by Lorentz, Maxwell's theory of light, Gauss's work on surfaces and Newton's law of gravitation and shows how they laid the groundwork for Einstein's revolutionary theory of relativity.

In addition to explaining the meaning of the general and special theories of relativity he also speculates on the far-reaching effect of Einstein's work upon science and the future of our modern world.

The noted science writer Leonard Engel called Bertrand Russell's ABC of Relativity *"one of the clearest and best accounts of relativity yet written."*

The Mysteries of Science from MENTOR

The ABC of Relativity

by
Bertrand Russell

Fourth Revised Edition
Edited by Felix Pirani

A MENTOR BOOK
NEW AMERICAN LIBRARY

This is an authorized reprint of a hardcover edition published by George Allen & Unwin, Ltd.

Library of Congress Catalog Card Number: 84-62504

 MENTOR TRADEMARK REG. U.S. PAT. OFF. AND FOREIGN COUNTRIES
REGISTERED TRADEMARK—MARCA REGISTRADA
HECHO EN CHICAGO, U.S.A.

SIGNET, SIGNET CLASSIC, MENTOR, PLUME, MERIDIAN and NAL BOOKS are published by New American Library, 1633 Broadway, New York, New York 10019

First Mentor Printing, May, 1959
Fourth Revised Edition (thirteenth printing), April, 1985

13 14 15 16 17 18 19 20 21

PRINTED IN THE UNITED STATES OF AMERICA

Note

This book first appeared in 1925. The basic principles of relativity have not changed since then, but both the theory and its applications have been much extended, and some revision has been necessary for the second and subsequent editions. For the second and third editions I carried out these revisions with Bertrand Russell's approval. The revisions for this fourth edition are entirely my responsibility. I have again altered a number of passages to agree with present knowledge or opinion, and I have attempted to eliminate the possessive case, as applied to laws or theories, where it seemed to me no longer appropriate.

I have also done my best to renounce the conventions that the masculine includes the feminine. Sixty years ago this may have been acceptable, or at least tolerated; now it is no longer so, and I have little doubt that Russell, who was a pro-feminist ahead of his time, would have approved of the renunciation.

I have not presumed to meddle with the substance of the last two chapters, which are largely philosophical rather than physical in character, although there is much in them with which I disagree.

F.P.

CONTENTS

CHAPTER 1

*Touch and Sight:
The Earth and the
Heavens*

Everybody knows that Einstein did something astonishing, but very few people know exactly what it was. It is generally recognized that he revolutionized our conception of the physical world, but the new conceptions are wrapped up in mathematical technicalities. It is true that there are innumerable popular accounts of the theory of relativity, but they generally cease to be intelligible just at the point where they begin to say something important. The authors are hardly to blame for this. Many of the new ideas can be expressed in nonmathematical language, but they are not less difficult on that account. What is demanded is a change in our imaginative picture of the world—a picture which has been handed down from remote, perhaps prehuman, ancestors and has been learned by each one of us in early childhood. A change in our imagination is always difficult, especially when we are no longer young. The same sort of change was demanded by Copernicus, who taught that the earth is not stationary and the heavens do not revolve about it once a day. To us now there is no difficulty in this idea, because we learned it before our mental habits had become fixed. Einstein's ideas, similarly, will seem easier to generations which grow up with them, but for us a certain effort of imaginative reconstruction is unavoidable.

In exploring the surface of the earth, we make use of all our senses, more particularly of the senses of touch and sight. In measuring lengths, parts of the human body are employed in prescientific ages: a "foot," a "cubit," a "span," are defined in this way. For longer distances, we think of the time it takes to walk from one place to another. We gradually learn to judge distance roughly by the eye, but we rely upon touch for accuracy. Moreover, it is touch that gives us our sense of "reality." Some things cannot be touched: rainbows, reflections in looking glasses, and so on. These things puzzle children, whose metaphysical speculations are arrested by the information that what is in the looking glass is not "real." Macbeth's dagger was unreal because it was not "sensible to feeling as to sight." Not only our geometry and physics, but our whole conception of what exists outside us, is based upon the sense of touch. We carry this even into our metaphors: a good speech is "solid," a bad speech is "gas," because we feel that a gas is not quite "real."

In studying the heavens, we are debarred from all senses except sight. We cannot touch the sun, or apply a foot rule to the Pleiades. Nevertheless, astronomers have unhesitatingly applied the geometry and physics which they found serviceable on the surface of the earth, and which they had based upon touch and travel. In doing so, they brought down trouble on their heads, which was not cleared up until relativity was discovered. It turned out that much of what had been learned from the sense of touch was unscientific prejudice, which must be rejected if we are to have a true picture of the world.

An illustration may help us to understand how much is impossible to the astronomer as compared with someone who is interested in things on the surface of the earth. Let us suppose that a drug is administered to you which makes you temporarily unconscious, and that when you wake you have lost your memory but not your reasoning powers. Let us suppose further that while you were unconscious you were carried into a balloon, which, when you come to, is sailing with the wind on a dark night—the night of the Fifth of November if you are in England, or of the Fourth of July if you are in America. You can see fireworks which are being sent off from the ground, from trains, and from planes traveling in all directions, but you cannot see the ground or the trains or the planes because of the darkness. What sort of picture of the world will you form? You will think that nothing is permanent: there are only brief flashes of light, which, during their short existence, travel through the void in the most various and bizarre curves. You cannot touch these flashes of light; you can only see them. Obviously your geometry and your physics and your metaphysics will be quite different from those of ordinary mortals. If ordinary mortals were with you in the balloon, you would find their speech unintelligible. But if Einstein were with you, you would understand him more easily than the ordinary mortal would, because you would be free from a host of preconceptions which prevent most people from understanding him.

The theory of relativity depends, to a considerable extent, upon getting rid of notions which are useful in ordinary life but not to our drugged balloonist. Circumstances on the surface of the earth, for various more or less accidental reasons, suggest conceptions which turn out to be inaccurate, although they have come to seem like necessities of thought. The most

Touch and Sight

important of these circumstances is that most objects on the earth's surface are fairly persistent and nearly stationary from a terrestrial point of view. If this were not the case, the idea of going on a journey would not seem so definite as it does. If you want to travel from King's Cross to Edinburgh, you know that you will find King's Cross where it has always been, that the railway line will take the course that it did when you last made the journey, and that Waverley Station in Edinburgh will not have walked up to the castle. You therefore say and think that you have traveled to Edinburgh, not that Edinburgh has traveled to you, though the latter statement would be just as accurate. The success of this common-sense point of view depends upon a number of things which are really of the nature of luck. Suppose all the houses in London were perpetually moving about, like a swarm of bees; suppose railways moved and changed their shapes like avalanches; and finally suppose that material objects were perpetually being formed and dissolved like clouds. There is nothing impossible in these suppositions. But obviously what we call a journey to Edinburgh would have no meaning in such a world. You would begin, no doubt, by asking the taxi driver: "Where is King's Cross this morning?" At the station you would have to ask a similar question about Edinburgh, but the reservations clerk would reply: "What part of

Edinburgh do you mean? Prince's Street has gone to Glasgow, the castle has moved up into the Highlands, and Waverley Station is under water in the middle of the Firth of Forth." And on the journey the stations would not be staying quiet, but some would be traveling north, some south, some east or west, perhaps much faster than the train. Under these conditions you could not say where you were at any moment. Indeed, the whole notion that one is always in some definite "place" is due to the fortunate immobility of most of the large objects on the earth's surface. The idea of "place" is only a rough practical approximation: there is nothing logically necessary about it, and it cannot be made precise.

If we were not much larger than an electron, we should not have this impression of stability, which is only due to the grossness of our senses. King's Cross, which to us looks solid, would be too vast to be conceived except by a few eccentric mathematicians. The bits of it that we could see would consist of little tiny points of matter, never coming into contact with each other, but perpetually whizzing around each other in an inconceivably rapid ballet dance. The world of our experience would be quite as mad as the one in which the different parts of Edinburgh go for walks in different directions. If—to take the opposite extreme—you were as large as the sun and lived as long, with a corresponding slowness of perception, you would again find a higgledy-piggledy universe without permanence: stars and planets would come and go like morning mists, and nothing would remain in a fixed position relatively to anything else. The notion of comparative stability which forms part of our ordinary outlook is thus due to the fact that we are about the size we are, and live on a planet of which the surface is not very hot. If this were not the case, we should not find prerelativity physics intellectually satisfying. Indeed, we should never have invented such theories. We should have had to arrive at relativity physics at one bound, or remain ignorant of scientific laws. It is fortunate for us that we were not faced with this alternative, since it is almost inconceivable that one person could have done the work of Euclid, Galileo, Newton, and Einstein. Yet without such an incredible genius physics could hardly have been discovered in a world where the universal flux was obvious to nonscientific observation.

In astronomy, although the sun, moon, and stars continue to exist year after year, yet in other respects the world we have to deal with is very different from that of everyday life. As already observed, we depend exclusively on sight: the heavenly bodies cannot be touched, heard, smelled, or tasted. Everything in the heavens is moving relatively to everything else. The earth is going around the sun; the sun is moving, very much faster than an express train, toward a point in the constellation Hercules; the "fixed" stars are scurrying hither and thither. There are no well-marked places in the sky, like King's Cross and Edinburgh. When you travel from place to place on the earth, you say the train moves and not the stations, because the stations preserve their topographical relations to each other and the surrounding country. But in astronomy it is arbitrary which you call the train and which the station: the question is to be decided purely by convenience and as a matter of convention.

In this respect, it is interesting to contrast Einstein and Copernicus. Before Copernicus, people thought that the earth stood still and the heavens revolved about it once a day. Copernicus taught that "really" the earth rotates once a day, and the daily revolution of sun and stars is only "apparent." Galileo and Newton endorsed this view, and many things were thought to prove it—for example, the flattening of the earth at the poles, and the fact that bodies are heavier there than at the equator. But in the modern theory the question between Copernicus and earlier astronomers is merely one of convenience; all motion is relative, and there is no difference between the two statements: "the earth rotates once a day" and "the heavens revolve about the earth once a day." The two mean exactly the same thing, just as it means

Touch and Sight

the same thing if I say that a certain length is six feet or two yards. Astronomy is easier if we take the sun as fixed than if we take the earth, just as accounts are easier in decimal coinage. But to say more for Copernicus is to assume absolute motion, which is a fiction. All motion is relative, and it is a mere convention to take one body as at rest. All such conventions are equally legitimate, though not all are equally convenient.

There is another matter of great importance in which astronomy differs from terrestrial physics because of its exclusive dependence upon sight. Both popular thought and old-fashioned physics used the notion of "force," which seemed intelligible because it was associated with familiar sensations. When we are walking, we have sensations connected with our muscles which we do not have when we are sitting still. In the days before mechanical traction, although people could travel by sitting in their carriages, they could see the horses exerting themselves, and evidently putting out "force" in the same way as human beings do. Everybody knew from experience what it is to push or pull, or to be pushed or pulled. These very familiar facts made "force" seem a natural basis for dynamics. But the Newtonian law of gravitation introduced a difficulty. The force between two billiard balls appeared intelligible because we know what it feels like to bump into another person but the force between the earth and the sun, which are ninety-three million miles apart, was mysterious. Even Newton regarded this "action at a distance" as impossible, and believed that there was some hitherto undiscovered mechanism by which the sun's influence was transmitted to the planets. However, no such mechanism was discovered, and gravitation remained a puzzle. The fact is that the whole conception of "gravitational force" is a mistake. The sun does not exert any force on the planets; in the relativity law of gravitation, the planet only pays attention to what it finds in its own neighbourhood. The way in which this works will be explained in a later chapter; for the present we are only concerned with the necessity of abandoning the notion of "gravitational force," which was due to misleading conceptions derived from the sense of touch.

As physics has advanced, it has appeared more and more that sight is less misleading than touch as a source of fundamental notions about matter. The apparent simplicity in the collision of billiard balls is quite illusory. As a matter of fact, the two billiard balls never touch at all; what really happens is inconceivably complicated, but is more analogous to what happens when a comet enters the solar system and goes away again than to what common sense supposes to happen.

Most of what we have said hitherto was al-

Touch and Sight

ready recognized by physicists before the theory of relativity was invented. It was generally held that motion is a merely relative phenomenon—that is to say, when two bodies are changing their relative position, we cannot say that one is moving while the other is at rest, since the occurrence is merely a change in their relation to each other. But a great labor was required in order to bring the actual procedure of physics into harmony with these new convictions. The technical methods of the old physics embodied the ideas of gravitational force and of absolute space and time. A new technique was needed, free from the old assumptions. For this to be possible, the old ideas of space and time had to be changed fundamentally. This is what makes both the difficulty and the interest of the theory. But before explaining it there are some preliminaries which are indispensable. These will occupy the next two chapters.

CHAPTER 2

What Happens and What Is Observed

A certain type of superior person is fond of asserting that "everything is relative." This is, of course, nonsense, because, if *everything* were relative, there would be nothing for it to be relative to. However, without falling into metaphysical absurdities it is possible to maintain that everything in the physical world is relative to an observer. This view, true or not, is *not* that adopted by the "theory of relativity." Perhaps the name is unfortunate; certainly it has led philosophers and uneducated people into confusions. They imagine that the new theory proves *everything* in the physical world to be relative, whereas, on the contrary, it is wholly concerned to exclude what is relative and arrive at a statement of physical laws that shall in no way depend upon the circumstances of the observer. It is true that these circumstances have been found to have more effect upon what appears to the observer than they were formerly thought to have, but at the same time the theory of relativity shows how to discount this effect completely. This is the source of almost everything that is surprising in the theory.

When two observers perceive what is regarded as one occurrence, there are certain similarities, and also certain differences, between their perceptions. The differences are obscured by the requirements of daily life, because from a practical point of view they are, as a rule, unimportant. But both psychology and physics, from their different angles, are compelled to emphasize the respects in which one person's perception of a given occurrence differs from another's. Some of these differences are due to differences in the brains or minds of the observers, some to differences in their sense organs, some to differences of physical situation: these three kinds may be called respectively psychological, physiological, and physical. A remark made in a language we know will be heard, whereas an equally loud remark in an unknown language may pass entirely unnoticed. Of two travelers in the Alps, one will perceive the beauty of the scenery while the other will notice the water-falls with a view to obtaining power from them. Such differences are psychological. The differences between a far-sighted and a short-sighted person or between a deaf person and someone who hears well, are physiological. Neither of these kinds concerns us, and I have mentioned them only in order to exclude them. The kind that concerns us is the purely physical kind. Physical differences between two observers will be preserved when the observers are replaced by cameras or recording machines, and can be reproduced in a film or on the phonograph. If two people both listen to a third person speaking, and one of them is nearer to the speaker than the other is, the nearer one will hear louder and slightly earlier sounds than are heard by the other. If two people both watch a tree falling, they see it from different angles. Both these differences would be shown equally by recording instruments: they are in no way due to idiosyncrasies in the observers, but are part of the ordinary course of physical nature as we experience it.

Physicists, like ordinary people, believe that their perceptions give them knowledge about what is really occurring in the physical world, and not only about their private experiences. Professionally, they regard the physical world as "real," not merely as something which human beings dream. An eclipse of the sun, for instance, can be observed by any person who is suitably situated, and is also observed by the photographic plates that are exposed for the purpose. The physicist is persuaded that something has really happened over and above the experience of those who have looked at the sun or at photographs of it. I have emphasized this point, which might seem a trifle obvious, because some people imagine that relativity made a difference in this respect. In fact, it has made none.

But if physicists are justified in this belief that a number of people can observe the "same" physical occurrence, then clearly physicists must be concerned with those features which the occurrence has in common for all observers, for the others cannot be regarded as belonging to the occurrence itself. At least, physicists must confine themselves to the features which are common to all "equally good" observers. Observers who use microscopes or telescopes are preferred to those who do not, because they see all that the latter see and more too. A sensitive photographic plate may "see" still more, and is then preferred to any eye. But such things as differences of perspective, or differences of apparent size, due to difference of distance, are obviously not attributable to the object; they belong solely to the point of view of the spectator. Common sense eliminates these in judging of objects; physics has to carry the same process much further, but the principle is the same.

I want to make it clear that I am not concerned with anything that can be called inaccuracy. I am concerned with genuine physical differences between occurrences each of which is a correct record of a certain event, from its own point of view. When a gun is fired, people who are not quite close to it see the flash before they hear the report. This is not due to any defect in their senses, but to the fact that sound travels more slowly than light. Light travels so fast that, from the point of view of most phenomena on the surface of the earth, it may be regarded as instantaneous. Anything that we can see on the earth happens practically at the moment when we see it. In a second, light travels 300,000 kilometers (about 186,000 miles). It travels from the sun to the earth in about eight minutes, and from the stars in anything from four years to several thousand million. Of course we cannot place a clock on the sun, send out a flash of light from it at 12 noon, Greenwich Mean Time, and have it received at Greenwich at 12:08 P.M. Our methods of estimating the speed of light are those we apply to sound when we use an echo. We can send a flash to a mirror and observe how long it takes for the reflection to reach us: this gives the time for the double journey to the mirror and back. If the distance to the mirror is measured, then the speed of light can be calculated.

Methods of measuring time are nowadays so precise that this procedure is used, not to calculate the speed of light, but to determine distances. By an international agreement, made in 1983, "the meter is the length of the path traveled in vacuum by light during a time 1/299 792 458 of a second." From the physicists' point of view, the speed of light has become a conversion factor, to be used for turning distances into times, just as the factor 0.9144 is used to turn distances in yards into distances in meters. It now makes perfectly good sense to say that the sun is about eight minutes away, or that it is a millionth of a second to the nearest bus stop.

The problem of allowing for the spectator's point of view, we may be told, is one of which physics has at all times been fully aware; indeed it has dominated astronomy ever since the time of Copernicus. This is true. But principles are often acknowledged long before their full consequences are drawn. Much of traditional physics is incompatible with the principle, in spite of the fact that it was acknowledged theoretically by all physicists.

There existed a set of rules which caused uneasiness to the philosophically minded, but were accepted by physicists because they worked in practice. Locke had distinguished "second-ary" qualities—colors, noises, tastes, smells, etc.—as subjective, while allowing "primary" qualities—shapes and positions and sizes—to be genuine properties of physical objects. The physicists' rules were such as would follow from this doctrine. Colors and noises were allowed to be subjective, but due to waves proceeding with a definite velocity—that of light or sound, as the case may be—from their source to the eye or ear of the percipient. Apparent shapes vary according to the laws of perspective, but these laws are simple and make it easy to infer the "real" shapes from several visual apparent shapes; moreover, the "real" shapes can be ascertained by touch in the case of bodies in our neighborhood. The objective time of a physical occurrence can be inferred from the time when we perceive it by allowing for the velocity of transmission—of light or sound or nerve currents according to circumstances. This was the view adopted by physicists in practice, whatever qualms they may have had in unprofessional moments.

This view worked well enough until physicists become concerned with much greater velocities than those that are common on the surface of the earth. An express train travels two miles in a minute; the planets travel a few miles in a second. Comets, when they are near the sun, travel much faster, but because of their continually changing shapes it is impossible to determine their positions very accurately. Practically, the planets were the most swiftly moving bodies to which dynamics could be adequately applied. With the discovery of radioactivity and cosmic rays, and recently with the construction of high-energy accelerating machines, new ranges of observation have become possible. Individual subatomic particles can be observed, moving with velocities not far short of that of light. The behavior of bodies moving with these enormous speeds is not what the old theories would lead us to expect. For one thing, mass seems to increase with speed in a perfectly definite manner. When an electron is moving very fast, a given force is found to have less effect upon it than when it is moving slowly. Then reasons have been found for thinking that the size of a body is affected by its motion—for example, if you take a cube and move it very fast, it gets shorter in the direction of its motion, from the point of view of a person who is not moving with it, though from its own point of view (i.e., for an observer traveling with it) it remains just as it was. What was still more astonishing was the discovery that lapse of time depends on motion; that is to say, two perfectly accurate clocks, one of which is moving very fast relatively to the other, will not continue to show the same time if they come together again after a journey. This is too small an effect to have been tested directly so far, but it should be possible to test it if we ever succeed in developing interstellar travel, for then we shall be able to make journeys long enough for this "time dilation," as it is called, to become quite appreciable.

There is some direct evidence for the time dilatation, but it is found in a different way. This evidence comes from observations of cosmic rays, which consist of a variety of atomic particles coming from outer space and moving very fast through the earth's atmosphere. Some of these particles, called mesons, disintegrate in flight, and the disintegration can be observed. It is found that the faster a meson is moving, the longer it takes to disintegrate, from the point of view of a scientist on the earth. It follows from results of this kind that what we discover by means of clocks and foot rules, which used to be regarded as the acme of impersonal science, is really in part dependent upon our private circumstances, i.e., upon the way in which we are moving relatively to the bodies measured.

This shows that we have to draw a different line from that which is customary in distinguishing between what belongs to the observer and what belongs to the occurrence which is being observed. If you put on blue spectacles, you know that the blue look of everything is due to the spectacles and does not belong to what you are looking at. But if you observe two flashes of lightning and note the interval of time between your observations: if you know where the flashes took place, and allow, in each case, for the time the light takes to reach you—in that case, if your chronometer is accurate, you may naturally think that you have discovered the actual interval of time between the two flashes, and not something merely personal to yourself. You will be confirmed in this view by the fact that all other careful observers to whom you have access agree with your estimates. This, however, is only due to the fact that all of you are on the earth and share its motion. Even two observers in spacecraft moving in opposite directions would have at the most a relative velocity of about 35,000 miles an hour, which is very little in comparison with 186,000 miles a second (the velocity of light). If an electron with a velocity of 170,000 miles a second could observe the time between the two flashes, it would arrive at a quite different estimate, after making full allowance for the velocity of light. How do you know this? the reader may ask. You are not an electron, you cannot move at these terrific speeds, no scientist has ever made the observations which would prove the truth of your assertion. Nevertheless, as we shall see in the sequel, there is good ground for the assertion—ground, first of all, in experiment, and—what is remarkable—ground in reasonings which could have been made at any time, but were not made until experiments had shown that the old reasonings must be wrong.

There is a general principle to which the theory of relativity appeals, which turns out to be more powerful than anybody would suppose. If you know that a person is twice as rich as another, this fact must appear equally whether you estimate the wealth of both in pounds or dollars or francs or any other currency. The numbers representing their fortunes will be changed, but one number will always be double the other. The same sort of thing, in more complicated forms, reappears in physics. Since all motion is relative, you may take any body you like as your standard body of reference, and estimate all other motions with reference to that one. If you are in a train and walking to the dining car, you naturally, for the moment, treat the train as fixed and estimate your motion in relation to it. But when you think of the journey you are making, you think of the earth as fixed, and say you are moving at the rate of sixty miles an hour. An astronomer who is concerned with the solar system takes the sun as fixed, and regards you as rotating and revolving; in comparison with this motion, that of the train is so slow that it hardly counts. An astronomer who is interested in the stellar universe may add the motion of the sun relatively to the average of the stars. You cannot say that one of these ways of estimating your motion is more correct than another; each is perfectly correct as soon as the reference body is assigned. Now just as you can estimate a person's fortune in different currencies without altering its relations to the fortunes of other people, so you can estimate a body's motion by means of different reference bodies without altering its relations to other motions. And as physics is entirely concerned with relations, it must be possible to express all the laws of physics by referring all motions to any given body as the standard.

We may put the matter in another way. Physics is intended to give information about what really occurs in the physical world, and not only about the private perceptions of separate observers. Physics must, therefore, be concerned with those features which a physical process has in common for all observers, since such features alone can be regarded as belonging to the physical occurrence itself. This requires that the *laws* of phenomena should be the same whether the phenomena are described as they appear to one observer or as they appear to another. This single principle is the generating motive of the whole theory of relativity.

Now what we have hitherto regarded as the spatial and temporal properties of physical occurrences are found to be in large part dependent upon the observer; only a residue can be attributed to the occurrences in themselves, and only this residue can be involved in the formulation of any physical law which is to have an a priori chance of being true. Einstein found ready to hand an instrument of pure mathematics, called the theory of tensors, which enabled him to discover laws expressed in terms of the objective residue and agreeing approximately with the old laws. Where the predictions of relativity theory differ from the old ones, they have hitherto proved more in accord with observation.

If there were no reality in the physical world, but only a number of dreams dreamed by different people, we should not expect to find any laws connecting the dreams of one person with the dreams of another. It is the close connection between the perceptions of one person and the (roughly) simultaneous perceptions of another that makes us believe in a common external origin of the different related perceptions. Physics accounts both for the likenesses and for the differences between different people's perceptions of what we call the "same" occurrence. But in order to do this it is first necessary for the physicist to find out just what are the likenesses.

What Happens and What Is Observed

They are not quite those traditionally assumed, because neither space nor time separately can be taken as strictly objective. What is objective is a kind of mixture of the two called "space-time." To explain this is not easy, but the attempt must be made; it will be begun in the next chapter.

CHAPTER 3

The Velocity of Light

Most of the curious things in the theory of relativity are connected with the velocity of light. The reader will be unable to grasp the reasons for such a serious theoretical reconstruction without some idea of the facts which made the old system break down.

The fact that light is transmitted with a definite velocity was first established by astronomical observations. Jupiter's moons are sometimes eclipsed by Jupiter, and it is easy to calculate the times when this ought to occur. It was found that when Jupiter was near the earth an eclipse of one of the moons would be observed a few minutes earlier than was expected; and when Jupiter was remote, a few minutes later than was expected. It was found that these deviations could all be accounted for by assuming that light has a certain velocity, so that what we observe to be happening in Jupiter really happened a little while ago—longer ago when Jupiter is distant than when it is near. Just the same velocity of light was found to account for similar facts in regard to other parts of the solar system. It was therefore accepted that light *in vacuo* always travels at a certain constant rate, almost exactly 300,000 kilometers a second. (A kilometer is about five-eighths of a mile.) When it became established that light consists of waves, this velocity was that of propagation of waves in the ether—at least they used to be in the ether, but now the ether has been given up, though the waves remain. This same velocity is that of radio waves (which are like light waves, only longer) and of X rays (which are like light waves, only shorter). It is generally held nowadays to be the velocity with which gravitation is propagated (before the discovery of relativity theory, it was thought that gravitation was propagated instantaneously, but this view is now untenable).

So far, all is plain sailing. But as it became possible to make more accurate measurements, difficulties began to accumulate. The waves were supposed to be in the ether, and therefore their velocity ought to be relative to the ether.

Now since the ether (if it exists) clearly offers no resistance to the motions of the heavenly bodies, it would seem natural to suppose that it does not share their motion. If the earth had to push a lot of ether before it, in the sort of way that a steamer pushes water before it, one would expect a resistance on the part of the ether analogous to that offered by the water to the steamer. Therefore, the general view was that the ether could pass through bodies without difficulty, like air through a coarse sieve, only more so. If this were the case, then the earth in its orbit must have a velocity relative to the ether. If, at some one point of its orbit, it happened to be moving exactly with the ether, it must at other points be moving through it all the faster. If you go for a circular walk on a windy day, you must be walking against the wind part of the way, whatever wind may be blowing; the principle in this case is the same. It follows that, if you choose two days six months apart, when the earth in its orbit is moving in exactly opposite directions, it must be moving against an ether wind on at least one of these days.

Now, if there is an ether wind, it is clear that, relatively to an observer on the earth, light signals will seem to travel faster with the wind than across it, and faster across it than against it. This is what Michelson and Morley set themselves to test by their famous experiment. They sent out light signals in two directions at right angles; each was reflected from a mirror, and came back to the place from which both had been sent out. Now anybody can verify, either by trial or by a little arithmetic, that it takes longer to row a given distance on a river up-stream and then back again, than it takes to row the same distance across the stream and back again. Therefore, if there were an ether wind, one of the two light signals, which consist of waves in the ether, ought to have traveled to the mirror and back at a slower average rate than the other. Michelson and Morley tried the experiment, they tried it in various positions, they tried it again later. Their apparatus was quite accurate enough to have detected the expected difference of speed or even a much smaller difference, if it had existed, but not the smallest difference could be observed. The result was a surprise to them as to everybody else, but careful repetitions made doubt impossible. The experiment was first made as long ago as 1881, and was repeated with more elaboration in 1887. But it was many years before it could be rightly interpreted.

The supposition that the earth carries the neighboring ether with it in its motion was found to be impossible, for a number of reasons. Consequently a logical deadlock seemed to have arisen, from which at first physicists sought to extricate themselves by very arbitrary hypotheses. The most important of these was that of Fitzgerald, developed by Lorentz, and now known as the Lorentz contraction hypothesis.

According to this hypothesis, when a body is in motion it becomes shortened in the direction of motion by a certain proportion depending upon its velocity. The amount of the contraction was to be just enough to account for the negative result of the Michelson-Morley experiment. The journey upstream and down again was to have been really a shorter journey than the one across the stream, and was to have been just so much shorter as would enable the slower light wave to traverse it in the same time. Of course, the shortening could never be detected by measurement, because our measuring rods would share it. A foot rule placed in the line of the earth's motion would be shorter than the same foot rule placed at right angles to the earth's motion. This point of view resembles nothing so much as the White Knight's "plan to dye one's whiskers green, and always use so large a fan that they could not be seen." The odd thing was that the plan worked well enough. Later on, when Einstein propounded the special theory of relativity (1905), it was found that the hypothesis was in a certain sense correct, but only in a certain sense. That is to say, the supposed contraction is not a physical fact, but a result of certain conventions of measurement which, when once the right point of view has been found, are seen to be such as we are almost compelled to adopt. But I do not wish yet to set forth Einstein's solution to the puzzle. For the present, it is the nature of the puzzle itself that I want to make clear.

On the face of it, and apart from hypotheses *ad hoc*, the Michelson-Morley experiment (in conjunction with others) showed that, relatively to the earth, the velocity of light is the same in all directions, and that this is equally true at all times of the year, although the direction of the earth's motion is always changing as it goes around the sun. Moreover, it appeared that this is not a peculiarity of the earth, but is true of all bodies: if a light signal is sent out from a body, that body will remain at the center of the waves as they travel outward, no matter how it may be moving—at least that will be the view of observers moving with the body. This was the plain and natural meaning of the experiments, and Einstein succeeded in inventing a theory which accepted it. But at first it was thought logically impossible to accept this plain and natural meaning.

A few illustrations will make it clear how very odd the facts are. When a shell is fired, it moves faster than sound: the people at whom it is fired first see the flash, then (if they are lucky) see the shell go by, and last of all hear the report. It is clear that if people could travel with the shell, they would never hear the report, as the shell would burst and kill them before the sound had overtaken them. But if sound worked on the same principles as light, they would hear everything just as if they were at rest. In that case, if a screen, suitable for producing echoes, were attached to the shell and traveling with it, say a hundred yards in front of it, they would hear the echo of the report from the screen after just the same interval of time as if they and the shell were at rest. This, of course, is an experiment which cannot be performed, but others which can be performed will show the difference. We might find some place on a railway where there is an echo from a place farther along the railway—say a place where the railway goes into a tunnel—and when a train is traveling along the railway, let someone on the bank fire a gun. If the train is traveling toward the echo, the passengers will hear the echo sooner than the person on the bank; if it is traveling in the opposite direction, they will hear it later. But these are not quite the circumstances of the Michelson-Morley experiment. The mirrors in that experiment correspond to the echo, and the mirrors are moving with the earth, so the echo ought to move with the train. Let us suppose that the shot is fired from the caboose, and the echo comes from a screen on the engine. We will suppose the distance from the caboose to the engine to be the distance that sound can travel in a second (about one-fifth of a mile), and the speed of the train to be one-twelfth of the speed of sound (about sixty miles an hour). We now have an experiment which can be performed by the people in the train. If the train were at rest, the guard would hear the echo in two seconds; as it is, it will take two and $^2/_{143}$ seconds. From this difference, knowing the velocity of sound, one can calculate the velocity of the train, even if it is a foggy night so that the banks are invisible. But if sound behaved like light, the echo would be heard by the guard after two seconds, however fast the train might be traveling.

Various other illustrations will help to show how extraordinary—from the point of view of tradition and common sense—are the facts about the velocity of light. Everyone knows that if you are on an escalator you reach the top sooner if you walk up than if you stand still. But if the escalator moved with the velocity of light (which it does not do even in New York), you would reach the top at exactly the same moment whether you walked up or stood still. Again: if you are walking along a road at the rate of four miles an hour, and a car passes you going in the same direction at the rate of forty miles an hour, if you and the car both keep going, the distance between you after an hour will be thirty-six miles. But if the car met you, going in the opposite direction, the distance after an hour would be forty-four miles. Now if the car were traveling with the velocity of light, it would make no difference whether it met or passed you: in either case, it would, after a second, be 186,000 miles from you. It would also be 186,000 miles from any other car which happened to be passing or meeting you less rapidly at the previous second. This seems impossible: how can the car be at the same distance from a number of different points along the road?

Let us take another illustration. When a fly touches the surface of a stagnant pool, it causes ripples which move outward in widening circles. The center of the circle at any moment is the point of the pool touched by the fly. If the fly moves about over the surface of the pool, it does not remain at the center of the ripples. But if the ripples were waves of light and the fly were a skilled physicist, it would find that it always remained at the center of the ripples, however it might move. Meanwhile, a skilled physicist sitting beside the pool would judge, as in the case of ordinary ripples, that the center was not the fly, but the point of the pool touched by the fly. And if another fly had touched the water at the same spot at the same moment, it also would find that it remained at the center of the ripples, even if it separated itself widely from the first fly. This is exactly analogous to the Michelson-Morley experiment. The pool corresponds to the ether; the fly corresponds to the earth; the contact of the fly and the pool corresponds to the light signal which Messrs. Michelson and Morley sent out; and the ripples correspond to the light waves.

Such a state of affairs seems, at first sight, quite impossible. It is no wonder that, although the Michelson-Morley experiment was made in 1881, it was not rightly interpreted until 1905. Let us see what, exactly, we have been saying. Take the example of the pedestrian and the car. Suppose there are a number of people at the same point of the road, some walking, some in cars; suppose they are going at varying rates, some in one direction and some in another. I say that if, at this moment, a light flash is sent out from the place where they all are, by each traveler's watch the light waves will be 186,000 miles from each one of them after a second, although the travelers will not any longer be all in the same place. At the end of a second by your watch it will be 186,000 miles from you, and it will also be 186,000 miles from any of the people who met you when it was sent out, after a second by their watches, even if they were moving in the opposite direction—assuming both to be perfect watches. How can this be?

There is only one way of explaining such facts, and that is to assume that watches and clocks are affected by motion. I do not mean that they are affected in ways that could be remedied by greater accuracy in construction; I mean something much more fundamental. I mean that if you say an hour has elapsed between two events, and if you base this assertion upon ideally careful measurements with ideally accurate chronometers, another equally precise person, who has been moving rapidly relatively to you, may judge that the time was more or less than an hour. You cannot say that one is right and the other wrong, any more than you could if one used a clock showing Greenwich time and another a clock showing New York time. How this comes about, I shall explain in the next chapter.

There are other curious things about the velocity of light. One is that no material body can ever travel as fast as light, however great may be the force to which it is exposed and however long the force may act. An illustration may help to make this clear. At exhibitions one sometimes sees a series of moving platforms, going round and round in a circle. The outside one goes at four miles an hour; the next goes four miles an hour faster than the first; and so on. You can step across from each to the next, until you find yourself going at a tremendous pace. Now you might think that, if the first platform does four miles an hour, and the second does four miles an hour relatively to the first, then the second does eight miles an hour relatively to the ground. This is an error; it does a little less, though so little less that not even the most careful measurements could detect the difference. I want to make quite clear what it is that I mean. Suppose that, in the morning, when the apparatus is just about to start, three couples each equipped with an ideally accurate chronometer stand in a row, one couple on the ground, one on the first platform, and one on the second. The first platform moves at the rate of four miles an hour with respect to the ground. Four miles an hour is 352 feet in a minute. The couple on the ground, after a minute by their watches, note the place on the ground opposite the couple on the first platform, who have been standing still while the platform carried them along. The couple on the ground measure the distance on the ground from themselves to the point opposite the couple on the first platform, and find it is 352 feet. The couple on the first platform, after a minute by their watches, note the point on their platform opposite to the couple on the second platform. The couple on the first platform measure the distance from themselves to the point opposite the couple on the second platform; it is again 352 feet. Problem: how far will the couple on the ground judge that the couple on the second platform have traveled in a minute? That is to say, if the couple on the ground, after a minute by their watches, note the place on the ground opposite the couple on the second platform, how far will this be from the couple on the ground? You would say, twice 352 feet—that is to say, 704 feet. But in fact it will be a little

less, though so little less as to be inappreciable.
The discrepancy results from the fact that the
two watches do not keep perfect time, in spite of
the fact that each is accurate from its owner's
point of view. If you had a long series of such
moving platforms, each moving four miles an
hour relatively to the one before it, you would
never reach a point where the last moving was
with the velocity of light relatively to the
ground, not even if you had millions of them.
The discrepancy, which is very small for small
velocities, becomes greater as the velocity increases, and makes the velocity of light an unat-
tainable limit. How all this happens, is the next
topic with which we must deal.

CHAPTER 4

Clocks and Foot Rules

Until the advent of the special theory of relativity, no one had thought that there could be any ambiguity in the statement that two events in different places happened at the same time. It might be admitted that, if the places were very far apart, there might be difficulty in finding out for certain whether the events were simultaneous, but everyone thought the meaning of the question perfectly definite. It turned out, however, that this was a mistake. Two events in distant places may appear simultaneous to one observer who has taken all due precautions to ensure accuracy (and, in particular, has allowed for the velocity of light), while another equally careful observer may judge that the first event preceded the second, and still another may judge that the second preceded the first. This would happen if the three observers were all moving rapidly relatively to each other. It would not be the case that one of them would be right and the other two wrong: they would all be equally right. The time order of events is in part dependent upon the observer; it is not always and altogether an intrinsic relation between the events themselves. The theory of relativity shows not only that this view accounts for the phenomena, but also that it is the one which ought to have resulted from careful reasoning based upon the old data. In actual fact, however, no one noticed the logical basis of the theory of relativity until the odd results of experiment had given a jog to people's reasoning powers.

How should we naturally decide whether two events in different places were simultaneous? One would naturally say: they are simultaneous if they are seen simultaneously by a person who is exactly halfway between them. (There is no difficulty about the simultaneity of two events in the *same* place, such, for example, as seeing a light and hearing a noise.) Suppose two flashes of lightning fall in two different places, say Greenwich Observatory and Kew Observatory. Suppose that St. Paul's is halfway between them, and that the flashes appear simultaneous to an observer on the dome of St. Paul's. In that case, a person at Kew will see the Kew flash first, and a person at Greenwich will see the Greenwich flash first, because of the time taken by light to travel over the intervening distance. But all three, if they are ideally accurate observers, will judge that the two flashes were simultaneous, because they will make the necessary allowance for the time of transmission of the light. (I am assuming a degree of accuracy far beyond human powers.) Thus, so far as observers on the earth are concerned, the definition of simultaneity will work well enough, so long as we are dealing with events on the surface of the earth. It gives results which are consistent with each other, and can be used for terrestrial physics in all problems in which we can ignore the fact that the earth moves.

But our definition is no longer so satisfactory when we have two sets of observers in rapid motion relatively to each other. Suppose we see what would happen if we substitute sound for light, and define two occurrences as simultaneous when they are heard simultaneously by someone halfway between them. This alters nothing in the principle, but makes the matter easier owing to the much slower velocity of sound. Let us suppose that on a foggy night two brigands shoot the guard and engine driver of a train. The guard is at the end of the train; the brigands are on the line and shoot their victims at close quarters. A passenger who is exactly in the middle of the train hears the two shots simultaneously. You would say, therefore, that the two shots were simultaneous. But a stationmaster who is exactly halfway between the two brigands hears the shot which kills the guard first. An Australian millionaire aunt of the guard and engine driver (who are cousins) has left her whole fortune to the guard, or, should he die first, to the engine driver. Vast sums are involved in the question who died first. The case goes to the House of Lords, and the lawyers on both sides, having been educated at Oxford, are agreed that either the passenger or the stationmaster must have been mistaken. In fact, both may perfectly well be right. The train travels away from the shot at the guard, and toward the shot at the engine driver; therefore, the noise of the shot at the guard has farther to go before reaching the passenger than the shot at the engine driver has. Therefore, if the passenger is right in saying that she heard the two reports simultaneously, the stationmaster must be right in saying that he heard the shot at the guard first.

We, who live on the earth, would naturally, in such a case, prefer the view of simultaneity obtained from a person at rest on the earth to the view of a person traveling in a train. But in theoretical physics no such parochial prejudices are permissible. A physicist on a comet, if there were one, would have just as good a right to a view of simultaneity as an earthly physicist has, but the results would differ, in just the same sort of way as in our illustration of the train and the shots. The train is not any more "real" in motion than the earth; there is no "really" about it. You might imagine a rabbit and a hippopotamus arguing as to whether people are "really" large animals; each would think its own point of view the natural one, and the other a pure flight of fancy. There is just as little substance in an argument as to whether the earth or the train is "really" in motion. And therefore, when we are defining simultaneity between distant events, we have no right to pick and choose among different bodies to be used in defining the point halfway between the events. All bodies have an equal right to be chosen. But if, for one body, the two events are simultaneous according to the definition, there will be other bodies for which the first precedes the second, and still others for which the second precedes the first. We cannot therefore say unambiguously that two events in distant places are simultaneous. Such a statement acquires a definite meaning only in relation to a definite observer. It belongs to the subjective part of our observation of physical phenomena, not to the objective part which is to enter into physical laws.

This question of time in different places is perhaps, for the imagination, the most difficult aspect of the theory of relativity. We are accustomed to the idea that everything can be dated. Historians make use of the fact that there was an eclipse of the sun visible in China on August 29, in the year 776 B.C.* No doubt astronomers could tell the exact hour and minute when the eclipse began to be total at any given spot in North China. And it seems obvious that we can speak of the positions of the planets at a given instant. The Newtonian theory enables us to calculate the distance between the earth and (say) Jupiter at a given time by the Greenwich clocks; this enables us to know how long light takes at that time to travel from Jupiter to the earth—say half an hour; this enables us to infer that half an hour ago Jupiter was where we see it now. All this seems obvious. But in fact it only works in practice because the relative velocities of the planets are very small compared with the velocity of light. When you judge that an event on the earth and an event on Jupiter have happened at the same time—for example, that Jupiter eclipsed one of its moons when the Greenwich clocks showed twelve midnight—a person moving rapidly relatively to the earth

---

* A contemporary Chinese ode, after giving the day of the year correctly, proceeds:

> For the moon to be eclipsed
> Is but an ordinary matter.
> Now that the sun had been eclipsed
> How bad it is!

would judge differently, assuming that both had made the proper allowance for the velocity of light. And naturally the disagreement about simultaneity involves a disagreement about periods of time. If we judged that two events on Jupiter were separated by twenty-four hours, another person moving rapidly relatively to Jupiter and the earth, might judge that they were separated by a longer time.

The universal cosmic time which used to be taken for granted is thus no longer admissible. For each body, there is a definite time order for the events in its neighborhood; this may be called the "proper" time for that body. Our own experience is governed by the proper time for our own body. As we all remain very nearly stationary on the earth, the proper times of different human beings agree, and can be lumped together as terrestrial time. But this is only the time appropriate to *large* bodies on the earth. For electrons in laboratories, quite different times would be wanted; it is because we insist upon using our own time that these particles seem to increase in mass with rapid motion. From their own point of view, their mass remains constant, and it is we who suddenly grow thin or corpulent. The history of a physicist as observed by an electron would resemble Gulliver's travels.

The question now arises: what really is measured by a clock? When we speak of a clock in the theory of relativity, we do not mean only clocks made by human hands: we mean anything which goes through some regular periodic performance. The earth is a clock, because it rotates once in every twenty-three hours and fifty-six minutes. An atom is a clock, because it emits light waves of very definite frequencies; these are visible as bright lines in the spectrum of the atom. The world is full of periodic occurrences, and fundamental mechanisms, such as atoms, show an extraordinary similarity in different parts of the universe. Any one of these periodic occurrences may be used for measuring time; the only advantage of humanly manufactured clocks is that they are especially easy to observe. However, some of the others are more accurate. Nowadays the standard of time is based on the frequency of a particular oscillation of cesium atoms, which is much more uniform than one based on the earth's rotation. But the question remains: if cosmic time is abandoned, what is really measured by a clock in the wide sense that we have just given to the term?

Each clock gives a correct measure of its own "proper" time, which, as we shall see presently, is an important physical quantity. But it does not give an accurate measure of any physical quantity connected with events on bodies that are moving rapidly in relation to it. It gives one

Clocks and Foot Rules

datum toward the discovery of a physical quantity connected with such events, but another datum is required, and this has to be derived from measurement of distances in space. Distances in space, like periods of time, are in general not objective physical facts, but partly dependent upon the observer. How this comes about must now be explained.

First of all, we have to think of the distance between two events, not between two bodies. This follows at once from what we have found as regards time. If two bodies are moving relatively to each other—and this is really always the case—the distance between them will be continually changing, so that we can only speak of the distance between them at a given time. If you are in a train traveling toward Edinburgh, we can speak of your distance from Edinburgh at a given time. But, as we said, different observers will judge differently as to what is the "same" time for an event in the train and an event in Edinburgh. This makes the measurement of distances relative, in just the same way as the measurement of times has been found to be relative. We commonly think that there are two separate kinds of interval between two events, an interval in space and an interval in time: between your departure from London and your arrival in Edinburgh, there are four hundred miles and ten hours. We have already seen that other observers will judge the time differently; it is even more obvious that they will judge the distance differently. An observer on the sun will think the motion of the train quite trivial, and will judge that you have traveled the distance traveled by the earth in its orbit and its diurnal rotation. On the other hand, a flea in the railway carriage will judge that you have not moved at all in space, but have afforded it a period of pleasure which it will measure by its "proper" time, not by Greenwich Observatory. It cannot be said that you or the sun dweller or the flea are mistaken: each is equally justified and is only wrong to ascribe an objective validity to subjective measures. The distance in space between two events is, therefore, not in itself a physical fact. But, as we shall see, there is a physical fact which can be inferred from the distance in time together with the distance in space. This is what is called the "interval" in space-time.

Taking any two events in the universe, there are two different possibilities as to the relation between them. It may be physically possible for a body to travel so as to be present at both events or it may not. This depends upon the fact that no body can travel as fast as light. Suppose, for example, that a flash of light is sent from the earth and reflected back from the moon. The time between the sending of the flash and the return of the reflection will be about two and a half seconds. No body could travel so fast as to be present on the earth during any part of those two and a half seconds and also present on the moon at the moment of the arrival of the flash, because in order to do so the body would have to travel faster than light. But theoretically a body could be present on the earth at any time before or after those two and a half seconds and also present on the moon at the time when the flash arrived. When it is physically impossible for a body to travel so as to be present at both events, we shall say that the interval* between the two events is "spacelike"; when it is physically possible for a body to be present at both events, we shall say that the interval between the two events is "timelike." When the interval is spacelike, it is possible for a body to move in such a way that an observer on the body will judge the two events to be simultaneous. In that case, the interval between the two events is what such an observer will judge to be the distance in space between them. When the interval is timelike, a body can be present at both events; in that case, the interval between the two events is what an observer on the body will judge to be the time between them, that is to say, it is "proper" time between the two events. There is a limiting case between the two, when the two events are parts of one light

---

*I shall define "interval" in a moment.

---

flash—or, as we might say, when the one event is the seeing of the other. In that case, the interval between the two events is zero.

There are thus three cases. (1) It may be possible for a ray of light to be present at both events; this happens whenever one of them is the seeing of the other. In this case the interval between the two events is zero. (2) It may happen that no body can travel from one event to the other, because in order to do so it would have to travel faster than light. In that case, it is always physically possible for a body to travel in such a way that an observer on the body would judge the two events to be simultaneous. The interval is what the observer would judge to be the distance in space between the two events. Such an interval is called "spacelike." (3) It may be physically possible for a body to travel so as to be present at both events; in that case, the interval between them is what an observer on such a body will judge to be the time between them. Such an interval is called "timelike."

The interval between two events is a physical fact about them, not dependent upon the particular circumstances of the observer.

There are two forms of the theory of relativity, the special and the general. The former is in general only approximate, but becomes very nearly exact at great distances from gravitating matter. Whenever gravitation may be neglected, the special theory can be applied, and then the interval between two events can be calculated when we know the distance in space and the distance in time between them, estimated by any observer. If the distance in space is greater than the distance that light would have traveled in the time, the separation is spacelike. Then the following construction gives the interval between the two events: draw a line AB as long as the distance that light would travel in the time; around A describe a circle whose radius is the distance in space between the two events; through B draw BC perpendicular to AB, meeting the circle in C. Then BC is the length of the interval between the two events.

When the distance is timelike, use the same figure, but let AC be now the distance that light would travel in the time, while AB is the distance in space between the two events. The interval between them is now the time that light would take to travel the distance BC.

Although AB and AC are different for different observers, BC is the same length for all observers, subject to corrections made by the general theory. It represents the one interval in "space-time" which replaces the two intervals in space and time of the older physics. So far, this notion of interval may appear somewhat mysterious, but as we proceed, it will grow less so, and its reason in the nature of things will gradually emerge.

CHAPTER 5

Space-time

Everybody who has ever heard of relativity knows the phrase "space-time," and knows that the correct thing is to use this phrase when formerly we should have said "space *and* time." But very few people who are not mathematicians have any clear idea of what is meant by this change of phraseology. Before dealing further with the special theory of relativity, I want to try to convey to the reader what is involved in the new phrase "space-time," because that is, from a philosophical and imaginative point of view, perhaps the most important of all the novelties that Einstein introduced.

Suppose you wish to say where and when

some event has occurred—say an explosion on an airplane—you will have to mention four quantities, say the latitude and longitude, the height above the ground, and the time. According to the traditional view, the first three of these give the position in space, while the fourth gives the position in time. The three quantities that give the position in space may be assigned in all sorts of ways. You might, for instance, take the plane of the equator, the plane of the meridian of Greenwich, and the plane of the 90th meridian, and say how far the airplane was from each of these planes; these three distances would be what are called "Cartesian coordinates," after Descartes. You might take any other three planes all at right angles to each other, and you would still have Cartesian coordinates. Or you might take the distance from London to a point vertically below the airplane, the direction of this distance (northeast, west-southwest, or whatever it might be), and the height of the airship above the ground. There are an infinite number of such ways of fixing the position in space, all equally legitimate; the choice between them is merely one of convenience.

When people said that space had three dimensions, they meant just this: that three quantities were necessary in order to specify the position of a point in space, but that the method of assigning these quantities was wholly arbitrary.

With regard to time, the matter was thought to be quite different. The only arbitrary elements in the reckoning of time were the unit, and the point of time from which the reckoning started. One could reckon in Greenwich time, or in Paris time, or in New York time; that made a difference as to the point of departure. One could reckon in seconds, minutes, hours, days, or years; that was a difference of unit. Both these were obvious and trivial matters. There was nothing corresponding to the liberty of choice as to the method of fixing position in space. And, in particular, it was thought that the method of fixing position in space and the method of fixing position in time could be made wholly independent of each other. For these reasons, people regarded time and space as quite distinct.

The theory of relativity has changed this. There are now a number of different ways of fixing position in time, which do not differ merely as to the unit and the starting point. Indeed, as we have seen, if one event is simultaneous with another in one reckoning, it will precede it in another, and follow it in a third. Moreover, the space and time reckonings are no longer independent of each other. If you alter the way of reckoning position in space, you may also alter the time interval between two events. If you alter the way of reckoning time, you may also alter the distance in space between two events. Thus space and time are no longer independent, any more than the three dimensions of space are. We still need four quantities to determine the position of an event, but we cannot, as before, divide off one of the four as quite independent of the other three.

It is not quite true to say that there is no longer any distinction between time and space. As we have seen, there are timelike intervals and spacelike intervals. But the distinction is of a different sort from that which was formerly assumed. There is no longer a universal time which can be applied without ambiguity to any part of the universe; there are only the various "proper" times of the various bodies in the universe, which agree approximately for two bodies which are not in rapid motion, but never agree exactly except for two bodies which are at rest relatively to each other.

The picture of the world which is required for this new state of affairs is as follows: suppose an event E occurs to me, and simultaneously a flash of light goes out from me in all directions. Anything that happens to any body after the light from the flash has reached it is definitely after the event E in any system of reckoning time. Any event anywhere which I could have seen before the event E occurred to me is definitely before the event E in any system of reck-

Space-time

oning time. But any event which happened in the intervening time is not definitely either before or after the event E. To make the matter definite: suppose I could observe a person in Sirius, and Sirian could observe me. Anything which the Sirian does, and which I see before the event E occurs to me, is definitely before E; anything the Sirian does after seeing the event E is definitely after E. But anything that the Sirian does before seeing the event E, which I see after the event E has happened, is not definitely before or after E. Since light takes about 8½ years to travel from Sirius to the earth, this gives a period of about 17 years in Sirius which may be called "contemporary" with E, since these years are not definitely before or after E.

Dr. A. A. Robb, in his *Theory of Time and Space,* suggested a point of view which may or may not be philosophically fundamental, but is at any rate a help in understanding the state of affairs we have been describing. He maintained that one event can only be said to be definitely *before* another if it can influence that other in some way. Now influences spread from a center at varying rates. Newspapers exercise an influence emanating from London at an average rate of about twenty miles an hour—rather more for long distances. Anything a person does on account of reading a newspaper article is clearly subsequent to the printing of the newspaper. Sounds travel much faster: it would be possible to arrange a series of loudspeakers along the main roads, and have newspapers shouted from each to the next. But telegraphing is quicker, and radio signals travel with the velocity of light, so that nothing quicker can ever be hoped for. Now what someone does in consequence of receiving a radio message is done *after* the message was sent; the meaning here is quite independent of conventions as to the measurement of time. But anything that is done while the message is on its way cannot be influenced by the sending of the message, and cannot influence the sender until some little time after the sending of the message, that is to say, if two bodies are widely separated, neither can influence the other except after a certain lapse of time; what happens before that time has elapsed cannot affect the distant body. Suppose, for instance, that some notable event happens on the sun: there is a period of sixteen minutes on the earth during which no event on the earth can have influenced or been influenced by the said notable event on the sun. This gives a substantial ground for regarding that period of sixteen minutes on the earth as neither before nor after the event on the sun.

The paradoxes of the special theory of relativity are only paradoxes because we are unaccustomed to the point of view, and in the habit of taking things for granted when we have no right to do so. This is especially true as regards

Space-time

the measurement of lengths. In daily life, our way of measuring lengths is to apply a foot rule or some other measure. At the moment when the foot rule is applied, it is at rest relatively to the body which is being measured. Consequently the length that we arrive at by measurement is the "proper" length, that is to say, the length as estimated by an observer who shares the motion of the body. We never, in ordinary life, have to tackle the problem of measuring a body which is in continual motion. And even if we did, the velocities of visible bodies on the earth are so small relatively to the earth that the anomalies dealt with by the theory of relativity would not appear. But in astronomy, or in the investigation of atomic structure, we are faced with problems which cannot be tackled in this way. Not being Joshua, we cannot make the sun stand still while we measure it; if we are to estimate its size we must do so while it is in motion relatively to us. And similarly if you want to estimate the size of an electron, you have to do so while it is in rapid motion, because it never stands still for a moment. This is the sort of problem with which the theory of relativity is concerned. Measurement with a foot rule, when it is possible, gives always the same result, because it gives the "proper" length of a body. But when this method is not possible, we find that curious things happen, particularly if the body to be measured is moving very fast relatively to the observer. A figure like the one at the end of the previous chapter will help us to understand the state of affairs.

Let us suppose that the body on which we wish to measure lengths is moving relatively to ourselves, and that in one second it moves the distance OM. Let us draw a circle around O whose radius is the distance that light travels in a second. Through M draw MP perpendicular to OM, meeting the circle in P. Thus OP is the distance that light travels in a second. The ratio of OP to OM is the ratio of the velocity of light to the velocity of the body. The ratio of OP to MP is the ratio in which apparent lengths are altered by the motion. That is to say, if the observer judges that two points in the line of motion on the moving body are at a distance from each other represented by MP, a person moving with the body would judge that they were at a distance represented (on the same scale) by OP. Distances on the moving body at right angles to the line of motion are not affected by the motion. The whole thing is reciprocal; that is to say, if an observer moving with the body were to measure lengths on the previous observer's body, they would be altered in just the same proportion. When two bodies are moving relatively to each other, lengths on either appear shorter to the other than to themselves. This is the Lorentz contraction, which was first invented to account for the result of the Michelson-Morley experiment. But it now emerges naturally from the fact that the two observers do not make the same judgment of simultaneity.

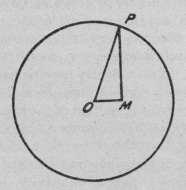

The way in which simultaneity comes in is this: we say that two points on a body are a foot apart when we can *simultaneously* apply one end of a foot rule to the one and the other end to the other. If, now, two people disagree about simultaneity, and the body is in motion, they will obviously get different results from their measurements. Thus the trouble about time is at the bottom of the trouble about distance.

The ratio of OP to MP is the essential thing in all these matters. Times and lengths and masses are all altered in this proportion when the body concerned is in motion relatively to the observer. It will be seen that, if OM is very much smaller than OP—that is to say, if the body is moving very much more slowly than light—MP and OP are very nearly equal, so that the alterations produced by the motion are very small. But if OM is nearly as large as OP—that is to say, if the body is moving nearly as fast as light—MP becomes very small compared to OP, and the effects become very great. The apparent increase of mass in swiftly moving particles had been observed, and the right formula had been found, before the invention of the special theory of relativity. In fact, Lorentz had arrived at the formulas called the Lorentz transformation, which embody the whole mathematical essence of the special theory of relativity. But it was Einstein who showed that the whole thing was what we ought to have expected, and not a set of makeshift devices to account for surprising experimental results. Nevertheless, it must not be forgotten that experimental results were the original motive of the whole theory, and have remained the ground for undertaking the tremendous logical reconstruction involved in the theory of relativity.

We may now recapitulate the reasons which have made it necessary to substitute "space-time" for space and time. The old separation of space and time rested upon the belief that there was no ambiguity in saying that two events in distant places happened at the same time; consequently, it was thought that we could describe the topography of the universe at a given instant in purely spatial terms. But now that simultaneity has become relative to a particular observer, this is no longer possible. What is, for one observer, a description of the state of the world at a given instant is, for another observer, a series of events at various different times, whose relations are not merely spatial but also temporal. For the same reason, we are concerned with *events*, rather than with *bodies*. In the old theory, it was possible to consider a number of bodies all at the same instant, and since the time was the same for all of them, it could be ignored. But now we cannot do that if we are to obtain an objective account of physical occurrences. We must mention the date at which a body is to be considered, and thus we arrive at an "event," that is to say, something which happens at a given time. When we know the time and place of an event in one observer's system of reckoning, we can calculate its time and place according to another observer. But we must know the time as well as the place, because we can no longer ask what is its place for the new observer at the "same" time as for the old observer. There is no such thing as the "same" time for different observers, unless they are at rest relatively to each other. We need four measurements to fix a position, and four measurements fix the position of an event in space-time, not merely of a body in space. Three measurements are not enough to fix any position. That is the essence of what is meant by the substitution of space-time for space and time.

CHAPTER 6

The Special Theory of Relativity

The special theory of relativity arose as a way of accounting for the facts of electromagnetism. We have here a somewhat curious history. In the eighteenth and early nineteenth centuries, the theory of electricity was wholly dominated by the Newtonian analogy. Two electric charges attract each other if they are of different kinds, one positive and one negative, but repel each other if they are of the same kind; in each case, the force varies as the inverse square of the distance, as in the case of gravitation. This force was conceived as an action at a distance, until Faraday, by a number of remarkable experiments, demonstrated the effect of the intervening medium. Faraday was no mathematician; Clerk Maxwell first gave a mathematical form to the results suggested by Faraday's experiments. Moreover, Clerk Maxwell gave grounds for thinking that light is an electromagnetic phenomenon, consisting of electromagnetic waves. The medium for the transmission of electromagnetic effects could therefore be taken to be the ether, which had long been assumed for the transmission of light. The correctness of Maxwell's theory of light was proved by the experiments of Hertz in manufacturing electromagnetic waves; these experiments afford the basis for radio and radar. So far, we have a record of triumphant progress, in which theory and experiment alternately assume the leading role. At the time of Hertz's experiments, the ether seemed securely established, and in just as strong a position as any other scientific hypothesis not capable of direct verification. But a new set of facts began to be discovered, and gradually the whole picture was changed.

The movement which culminated with Hertz was a movement for making everything continuous. The ether was continuous, the waves in it were continuous, and it was hoped that matter would be found to consist of some continuous structure in the ether. But then came the discovery of the atomic structure of matter, and of the discrete structure of the atoms themselves.

Atoms were believed to be built up of electrons, protons, and neutrons. The electron is a small particle bearing a definite charge of negative electricity. The proton bears a definite charge of positive electricity, while the neutron is not charged. (It is only a matter of custom that the charge on the electron is called negative and the charge on the proton positive, rather than the other way around.) It appeared probable that electricity was not to be found except in the form of the charges on the electron and proton; all electrons have exactly the same negative charge, and all protons have an exactly equal and opposite positive charge. Later on, other subatomic particles were discovered; most of them are called mesons or hyperons. All protons have exactly the same weight; they are about eighteen hundred times as heavy as electrons. All neutrons also have exactly the same weight; they are slightly heavier than protons. Mesons, of which there are several different kinds, weigh more than electrons but less than protons, while hyperons are heavier than protons or neutrons.

Some of the particles bear electric charges, while others do not. It is found that all the positively charged ones have exactly the same charge as the proton, while all the negatively charged ones have exactly the same charge as the electron, although their other properties are quite different. To confuse matters, there is a particle which is identical with the electron, except that it has a positive charge instead of a negative one; it is called the positron. It is possible to manufacture experimentally a particle which is identical with the proton except that it has a negative charge; it is called the antiproton.

These discoveries about the discrete structure of matter are inseparable from the discoveries of other so-called quantum phenomena, like the bright lines in the spectrum of an atom. It seems that all natural processes show a fundamental discontinuity whenever they can be measured with sufficient precision.

Thus physics has had to digest new facts and face new problems. Although the quantum theory has existed in more or less its present form for sixty years, and the special theory of relativity for eighty, little progress was made, until about thirty years ago, in connecting the two together. Recent developments in the quantum theory have made it more consistent with special relativity, and these improvements have helped our understanding of the subatomic particles a good deal, but many serious difficulties remain.

The problems solved by the special theory of relativity in its own right, quite apart from the quantum theory, are typified by the Michelson-Morley experiment. Assuming the correctness

The Special Theory of Relativity

of Maxwell's theory of electromagnetism there should have been certain discoverable effects of motion through the ether; in fact, there were none. Then there was the observed fact that a body in very rapid motion appears to increase its mass; the increase is in the ratio of OP to MP in the figure in the preceding chapter. Facts of this sort gradually accumulated until it became imperative to find some theory which would account for them all.

Maxwell's theory reduced itself to certain equations, known as Maxwell's equations. Through all the revolutions which physics has undergone in the last century, these equations have remained standing; indeed, they have continually grown in importance as well as in certainty—for Maxwell's arguments in their favor were so shaky that the correctness of his results must almost be ascribed to intuition. Now these equations were, of course, founded upon experiments in terrestrial laboratories, but there was a tacit assumption that the motion of the earth through the ether could be ignored. In certain cases, such as the Michelson-Morley experiment, this ought not to have been possible without measurable error, but it turned out to be always possible. Physicists were faced with the odd difficulty that Maxwell's equations were more accurate than they should be. A very similar difficulty was explained by Galileo at the very beginning of modern physics. Most people think that if you let a weight drop it will fall vertically. But if you try the experiment in the cabin of a moving ship, the weight falls, in relation to the cabin, just as if the ship were at rest; for instance, if it starts from the middle of the ceiling it will drop onto the middle of the floor. That is to say, from the point of view of an observer on the shore it does not fall vertically, since it shares the motion of the ship. So long as the ship's motion is steady, everything goes on inside the ship as if the ship were not moving. Galileo explained how this happens, to the great indignation of the disciples of Aristotle. In orthodox physics, which is derived from Galileo, a uniform motion in a straight line has no discoverable effects. This was, in its day, as astonishing a form of relativity as that of Einstein is to us. Einstein, in the special theory of relativity, set to work to show how electromagnetic phenomena could be unaffected by uniform motion through the ether—if there be an ether. This was a more difficult problem, which could not be solved by merely adhering to the principles of Galileo.

The really difficult effort required for solving this problem was in regard to time. It was necessary to introduce the notion of "proper" time which we have already considered, and to abandon the old belief in one universal time. The quantitative laws of electromagnetic phenomena are expressed in Maxwell's equations and these equations are found to be true for all observers, however they may be moving. It is a straightforward mathematical problem to find out what differences there must be between the measures applied by one observer and the measures applied by another, if, in spite of their relative motion, they are to find the same equations verified. The answer is contained in the Lorentz transformation, found as a formula by Lorentz, but interpreted and made intelligible by Einstein.

The Lorentz transformation tells us what estimate of distances and periods of time will be made by an observer whose relative motion is known, when we are given those of another observer. We may suppose that you are in a train on a railway which travels due east. You have been traveling for a time which, by the clocks at the station from which you started, is t. At a distance x from your starting point, as measured by the people on the line, an event occurs at this moment—say the line is struck by lightning. You have been traveling all the time with a uniform velocity v. The question is: how far from you will you judge that this event has taken place, and how long after you started will it be by your watch, assuming that your watch is correct from the point of view of an observer on the train?

Our solution of this problem has to satisfy certain conditions. It has to bring out the result that the velocity of light is the same for all observers, however they may be moving. And it has to make physical phenomena—in particular, those of electromagnetism—obey the same laws for different observers, however they may find their measures of distances and times affected by their motion. And it has to make all such effects on measurement reciprocal. That is to say, if you are in a train and your motion affects your estimate of distances outside the train, there must be an exactly similar change in the estimate which people outside the train make of distances inside it. These conditions are sufficient to determine the solution of the problem, but the solution requires more mathematics than I have allowed myself in this book.

Before dealing with the matter in general terms, let us take an example. Let us suppose that you are in a train on a long straight railway, and that you are traveling due east at three-fifths of the velocity of light. Suppose that you measure the length of your train, and find that it is a hundred yards. Suppose that the people who catch a glimpse of you as you pass succeed, by skillful scientific methods, in taking observations which enable them to calculate the length of your train. If they do their work correctly, they will find that it is eighty yards long. Everything in the train will seem to them shorter in the direction of the train than it does to you. Dinner plates, which you see as ordinary circular plates, will look to the outsider as if they were oval: they will seem only four-fifths as broad in the direction in which the train is moving as in the direction of the breadth of the train. And all this is reciprocal. Suppose you see out of the window a fishing rod, carried by someone who measures it to be fifteen feet long. If it is held upright, you will also see it to be fifteen feet long; so you will if it is held horizontally at right angles to the railway. But if it is pointed along the railway, it will seem to you to be only twelve feet long. In describing what is seen, I have assumed that everyone makes due allowance for perspective. Despite this, all the lengths of objects in the train will be diminished by 20 percent, in the direction of motion, for people outside, and so will those of objects outside, for you in the train.

But the effects in regard to time are even stranger. This matter was explained with almost ideal lucidity by Eddington, and my example is based on one given by him:

Imagine a spacecraft which moves away from the earth at a speed of 156,000 miles a second. If you were able to observe the people in the spacecraft, you would infer that they were unusually slow in their movements, and other events in the vehicle would be similarly retarded. Everything which took place there would seem to take twice as long as usual. I say "infer" deliberately; you would see a still more extravagant slowing down of time, but that is easily explained, because the spacecraft is rapidly increasing its distance from you and the light impressions take longer and longer to reach you. The more moderate retardation referred to remains after you have allowed for the time of transmission of light. But here reciprocity comes in, because from the point of view of the space travelers you are moving away from them at 156,000 miles a second, and when they have made all allowances, they find that it is you who are sluggish.

This question of time is rather intricate, owing to the fact that events which one person judges to be simultaneous another considers to be separated by a lapse of time. In order to try to make clear how time is affected, I shall revert to our railway train traveling due east at a rate of three-fifths that of light. For the sake of illustration, I assume that the earth is large and flat, instead of small and round.

If we take events which happen at a fixed point on the earth, and ask ourselves how long after the beginning of the journey they will seem to be to the travelers, the answer is that there will be that retardation that Eddington speaks of, which means in this case that what seems an hour in the life of the people on the ground is judged to be an hour and a quarter by the travelers who observe them from the train. Reciprocally, what seems an hour in the lives of the people in the train is judged by the people observing from outside to be an hour and a quarter. Each makes periods of time observed in the lives of the others a quarter as long again as they are to those who live through them. The proportion is the same in regard to times as in regard to lengths.

But when, instead of comparing events at the same place on the earth, we compare events at widely separated places, the results are still more odd. Let us now take all the events along the railway, which from the point of view of people who are stationary on the earth, happen at a given instant, say the instant when the train passes a certain signal. Of these events, those which occur at points toward which the train is moving will seem to the travelers to have already happened, while those which occur at points behind the train will, for them, be still in the future. When I say that events in the forward direction will seem to have already happened, I am saying something not strictly accurate; because they will not yet have seen them, but when they do see them, they will, after allowing for the velocity of light, come to the conclusion that these events must have happened before the moment in question. An event which happens in the forward direction along the railway, and which the stationary observers judge to be now (or rather, will judge to have been now when they come to know of it) if it occurs at a distance along the line which light could travel in a second, will be judged by the travelers to have occurred three-quarters of a second ago. If it occurs at a distance which the people on the earth judge that light could travel in a year, the travelers will judge (when they come to know of it) that it occurred nine months earlier than the moment when they pass the earth dwellers. And generally, they will antedate events in the forward direction along the railway by three-quarters of the time that it would take light to travel from them to those on the earth whom they are just passing, and who hold that these events are happening now—or rather, will hold that they happened now when the light from the events reaches them. Events happening on the railway behind the train will be postdated by an exactly equal amount.

We have thus a twofold correction to make in the date of an event when we pass from the terrestrial observers to the travelers. We must first take five-fourths of the time as estimated by the earth dwellers, and then subtract three-fourths of the time that it would take light to travel from the event in question to the earth dwellers.

Take some event in a distant part of the universe, which becomes visible to the earth dwellers and the travelers just as they pass each other. The earth dwellers, if they know how far off the event occurred, can judge how long ago it occurred, since they know the speed of light. If the event occurred in the direction toward which the travelers are moving, the travelers will infer that it happened twice as long ago as the earth dwellers think. But if it occurred in the direction from which they have come, they will argue that it happened only half as long ago as the earth dwellers think. If the travelers move at a different speed, these proportions will be different.

Suppose now that (as sometimes occurs) two new stars have suddenly flared up and have just become visible to the travelers and to the earth dwellers whom they are passing. Let one of them be in the direction toward which the train is traveling, the other in the direction from which it has come. Suppose that the earth dwellers are able, in some way, to estimate the distance of the two stars, and to infer that light takes fifty years to reach them from the one in the direction toward which the travelers are moving, and one hundred years to reach them from the other. The earth dwellers will then argue that the explosion which produced the new star in the forward direction occurred fifty years ago, while the explosion which produced the other new star occurred a hundred years ago. The travelers will exactly reverse these figures: they will infer that the forward explosion occurred a hundred years ago, and the backward one fifty years ago. I assume that both groups argue correctly on correct physical data. In fact, both are right, unless they imagine that the other must be wrong. It should be noted that both will have the same estimate of the velocity of light, because their estimates of the distances of the two new stars will vary in exactly the same proportion as their estimates of the times since the explosions. Indeed, one of the main motives of this whole theory is to secure that the velocity of light shall be the same for all observers, however they may be moving. This fact, established by experiment, was incompatible with the old theories, and made it absolutely necessary to admit something startling. The theory of relativity is just as little startling as it is compatible with the facts. Indeed, after a time, it ceases to seem startling at all.

There is another feature of very great importance in the theory we have been considering, and that is that, although distances and times

The Special Theory of Relativity

vary for different observers, we can derive from them the quantity called "interval," which is the same for all observers. The "interval," in the special theory of relativity, is obtained as follows: take the square of the distance between two events, and the square of the distance traveled by light in the time between the two events; subtract the lesser of these from the greater, and the result is defined as the square of the interval between the events. The interval is the same for all observers and represents a genuine physical relation between the two events, which the time and the distance do not. We have already given a geometrical construction for the interval at the end of Chapter 4; this gives the same result as the above rule. The interval is "timelike" when the time between the events is longer than light would take to travel from the place of the one to the place of the other; in the contrary case it is "spacelike." When the time between the two events is exactly equal to the time taken by light to travel from one to the other, the interval is zero; the two events are then situated on parts of one light ray, unless no light happens to be passing that way.

When we come to the general theory of relativity, we shall have to generalize the notion of interval. The more deeply we enter into the structure of the world, the more important this concept becomes; we are tempted to say that it is the reality of which distances and periods of time are confused representations. The theory of relativity has altered our view of the fundamental structure of the world; that is the source both of its difficulty and of its importance.

The remainder of this chapter may be omitted by readers who have not even the most elementary acquaintance with geometry or algebra. But for the benefit of those whose education has not been *entirely* neglected, I will add a few explanations of the general formula of which I have hitherto given only particular examples. The general formula in question is the Lorentz transformation, which tells, when one body is moving in a given manner relatively to another, how to infer the measures of lengths and times appropriate to the one body from those appropriate to the other. Before giving the algebraical formula, I will give a geometrical construction. As before, we will suppose that there are two observers, whom we will call O and O', one of whom is stationary on the earth while the other is traveling at a uniform speed along a straight railway. At the beginning of the time considered, the two observers were at the same point of the railway, but now they are separated by a certain distance. A flash of lightning strikes a point X on the railway, and O judges that at the moment when the flash takes place the observer in the train has reached the point O'. The problem is: what is the distance from O' to the flash, and how long after the beginning of the journey (when O' and O were together) did it take place, as judged by O'? We are supposed to know O's estimates, and we want to calculate those of O'.

In the time that, according to O, has elapsed since the beginning of the journey, let OC be the distance that light would have traveled along the railway. Describe a circle about O, with OC as radius, and through O' draw a perpendicular to the railway, meeting the circle in D. On OD take a point Y such that OY is equal to OX (X is the point of the railway where the lightning strikes). Draw YM perpendicular to the railway, and OS perpendicular to OD. Let YM and OS meet in S. Also let DO' produced and OS produced meet in R. Through X and C draw perpendiculars to the railway meeting OS produced and in Q and Z respectively. Then RQ (as measured by O) is the distance from O' to the flash, as judged by O'. According to the old view, the distance would be O'X. And whereas O thinks that, in the time from the beginning of the journey to the flash, light would travel a distance OC, O' thinks that the time elapsed is that required for light to travel the distance SZ (as measured by O). The interval as measured by O is got by subtracting the square on OX from the square on OC; the interval as measured by O' is got by subtracting the square on RQ from the square on SZ. A little very elementary geometry shows that these are equal.

The algebraical formulas embodied in the above construction are as follows: from the point of view of O, let an event occur at a distance x along the railway, and at a time t after the beginning of the journey (when O' was at O). From the point of view of O', let the same event occur at a distance x' along the railway, and at a time t' after the beginning of the journey. Let c be the velocity of light, and v the velocity of O' relative to O. Put

$$\beta = \frac{c}{\sqrt{c^2 - v^2}}$$

Then

$$x' = \beta \, (x - v \, t)$$

$$t' = \beta \left(t - \frac{vx}{c^2} \right)$$

This is the Lorentz transformation, from which everything in this chapter can be deduced.

CHAPTER 7

Intervals in Space-time

The special theory of relativity, which we have been considering hitherto, solved completely a certain definite problem: to account for the experimental fact that, when two bodies are in uniform relative motion, all the laws of physics, both those of ordinary dynamics and those connected with electricity and magnetism, are exactly the same for the two bodies. "Uniform" motion, here, means motion in a straight line with constant velocity. But although one problem was solved by the special theory, another was immediately suggested: what if the motion of the two bodies is not uniform? Suppose, for instance, that one is the earth while the other is

Intervals in Space-time

a falling stone. The stone has an accelerated motion: it is continually falling faster and faster. Nothing in the special theory enables us to say that the laws of physical phenomena will be the same for an observer on the stone as for one on the earth. This is particularly awkward, as the earth itself is, in an extended sense, a falling body: it has at every moment an acceleration* toward the sun, which makes it go around the sun instead of moving in a straight line. As our knowledge of physics is derived from experiments on the earth, we cannot rest satisfied with a theory in which the observer is supposed to have no acceleration. The general theory of relativity removes this restriction and allows the observer to be moving in any way, straight or crooked, uniformly or with an acceleration. In the course of removing the restriction, Einstein was led to his new law of gravitation, which we shall consider presently. The work was extraordinarily difficult, and occupied him for ten years. The special theory dates from 1905, the general theory from 1915.

It is obvious from experiences with which we are all familiar that an accelerated motion is

---

*Not only an increase in speed, but any change in speed or direction, is called "acceleration." The only sort of motion called "unaccelerated" is motion with constant speed *in a straight line.*

much more difficult to deal with than a uniform one. When you are in a train which is traveling steadily, the motion is not noticeable so long as you do not look out of the window but when the brakes are applied suddenly you are precipitated forward, and you become aware that something is happening without having to notice anything outside the train. Similarly in an elevator everything seems ordinary while it is moving steadily, but at starting and stopping, when its motion is accelerated, you have odd sensations in the pit of the stomach. (We call a motion "accelerated" when it is getting slower as well as when it is getting quicker; when it is getting slower the acceleration is negative.) The same thing applies to dropping a weight in the cabin of a ship. So long as the ship is moving uniformly, the weight will behave, relatively to the cabin, just as if the ship were at rest: if it starts from the middle of the ceiling, it will hit the middle of the floor. But if there is an acceleration everything is changed. If the boat is increasing its speed very rapidly, the weight will seem to an observer in the cabin to fall in a curve directed toward the stern; if the speed is being rapidly diminished, the curve will be directed toward the bow. All these facts are familiar, and they led Galileo and Newton to regard an accelerated motion as something radically different, in its own nature, from a uniform motion. But this distinction could only be maintained by regarding motion as absolute, not relative. If all motion is relative, the earth is accelerated relatively to the lift just as truly as the lift relatively to the earth. Yet the people on the ground have no sensations in the pits of their stomachs when the elevator starts to go up. This illustrates the difficulty of our problem. In fact, though few physicists in modern times have believed in absolute motion, the technique of mathematical physics still embodied Newton's belief in it, and a revolution in method was required to obtain a technique free from this assumption. This revolution was accomplished in Einstein's general theory of relativity.

It is somewhat optional where we begin in explaining the new ideas which Einstein introduced, but perhaps we shall do best by taking the conception of "interval." This conception, as it appears in the special theory of relativity, is already a generalization of the traditional notion of distance in space and time, but it is necessary to generalize it still further. However, it is necessary first to explain a certain amount of history, and for this purpose we must go back as far as Pythagoras.

Pythagoras, like many of the greatest characters in history, perhaps never existed: he or she is a semimythical character, who combined mathematics and priestcraft in uncertain proportions. I shall, however, assume that

Pythagoras existed and discovered the theorem attributed to someone of this name. Pythagoras was roughly a contemporary of Confucius and Buddha who founded a religious sect and a school of mathematicians who took a particular interest in right-angled triangles. The theorem of Pythagoras (the 47th proposition of Euclid) states that the sum of the squares on the two shorter sides of a right-angled triangle is equal to the square on the side opposite the right angle. No proposition in the whole of mathematics has had such a distinguished history. We all learned to "prove" it in youth. It is true that the "proof" proved nothing, and that the only way to prove it is by experiment. It is also the case that the proposition is not *quite* true—it is only approximately true. But everything in geometry, and subsequently in physics, has been derived from it by successive generalizations. One of these generalizations is the general theory of relativity.

The theorem of Pythagoras was itself, in all probability, a generalization of an Egyptian rule of thumb. In Egypt, it had been known for ages that a triangle whose sides are 3, 4, and 5 units of length is a right-angled triangle; the Egyptians used this knowledge practically in measuring their fields. Now, if the sides of a triangle are 3, 4, and 5 inches, the squares on these sides will contain respectively 9, 16, and 25 square inches; and 9 and 16 added together make 25. Three times three is written 3^2; four times four, 4^2; five times five, 5^2. So that we have

$$3^2 + 4^2 = 5^2.$$

It is supposed that Pythagoras noticed this fact, after learning from the Egyptians that a triangle whose sides are 3, 4, and 5 has a right angle. Pythagoras found that this could be generalized, and so arrived at the famous theorem: in a right-angled triangle, the square on the side opposite the right angle is equal to the sum of the squares on the other two sides.

Similarly in three dimensions: if you take a right-angled solid block, the square on the diagonal (the dotted line in the figure) is equal to the sum of the squares on the three sides.

This is as far as the ancients got in this matter.

The next step of importance is due to Descartes, who made the theorem of Pythagoras the basis of the method of analytical geometry. Suppose you wish to map out systematically all the places on a plain—we will suppose it small enough to make it possible to ignore the fact that the earth is round. We will suppose that you live in the middle of the plain. One of the simplest ways of describing the position of a place is to say: starting from my house, go first such and such a distance east, then such and such a distance north (or it may be west in the first case, and south in the second). This tells

you exactly where the place is. In the rectangular cities of America, it is the natural method to adopt: in New York you will be told to go so many blocks east (or west) and then so many blocks north (or south). The distance you have to go east is called x, and the distance you have to go north is called y. (If you have to go west, x is negative; if you have to go south, y is negative.) Let O be your starting point (the "origin"); let OM be the distance you go east, and MP the distance you go north. How far are you from home in a direct line when you reach P? The theorem of Pythagoras gives the answer. The square on OP is the sum of the squares on OM and MP. If

OM is four miles and MP is three miles, OP is five miles. If OM is twelve miles and MP is five miles, OP is thirteen miles, because $12^2 + 5^2 = 13^2$. So that if you adopt Descartes' method of mapping, the theorem of Pythagoras is essential in giving you the distance from place to place. In three dimensions the thing is exactly analogous. Suppose that, instead of wanting merely to fix positions on the plain, you want to fix stations for captive balloons above it, you will then have to add a third quantity, the height at which the balloon is to be. If you call the height z, and if r is the direct distance from O to the balloon, you will have

$$r^2 = x^2 + y^2 + z^2,$$

and from this you can calculate r when you know x, y, and z. For example, if you can get to the balloon by going 12 miles east, 4 miles north, and then 3 miles up, your distance from the balloon in a straight line is thirteen miles, because $12 \times 12 = 144$, $4 \times 4 = 16$, $3 \times 3 = 9$, $144 + 16 + 9 = 169 = 13 \times 13$.

But now suppose that, instead of taking a small piece of the earth's surface which can be regarded as flat, you consider making a map of the world. An accurate map of the world on flat paper is impossible. A globe can be accurate, in the sense that everything is produced to scale, but a flat map cannot be. I am not talking of practical difficulties, I am talking of a theoretical impossibility. For example: the northern halves of the meridian of Greenwich and of the 90th meridian of west longitude, together with the piece of the equator between them, make a triangle whose sides are all equal and whose angles are all right angles. On a flat surface, a triangle of that sort would be impossible. On the other hand, it is possible to make a square on a flat surface, but on a sphere it is impossible. Suppose you try on the earth: walk 100 miles west, then 100 miles north, then 100 miles east, then 100 miles south. You might think this would make a square, but it wouldn't, because you would not at the end have come back to your starting point. If you have time, you may convince yourself of this by experiment. If not, you can easily see that it must be so. When you are nearer the pole, 100 miles takes you through more longitude than when you are nearer the equator, so that in doing your 100 miles east (if you are in the northern hemisphere) you get to a point farther east than that from which you started. As you walk due south after this, you remain farther east than your starting point, and end up at a different place from that in which you began. Suppose, to take another illustration, that you start on the equator 4,000 miles east of the Greenwich meridian; you travel till you reach the meridian, then you travel northward along it for 4,000 miles, through Greenwich and up to the neighborhood of the Shetland Islands; then you travel eastward for 4,000 miles, and then 4,000 miles south. This will take you to the equator to a point about 4,000 miles farther east than the point from which you started.

In a sense, what we have just been saying is not quite fair, because except on the equator, traveling due east is not the shortest route from a place to another place due east of it. A ship traveling (say) from New York to Lisbon, which is nearly due east, will start by going a certain distance northward. It will sail on a "great circle," that is to say, a circle whose center is the center of the earth. This is the nearest approach to a straight line that can be drawn on the surface of the earth. Meridians of longitude are great circles, and so is the equator, but the other parallels of latitude are not. We ought, therefore, to have supposed that, when you reach the Shetlands Islands, you travel 4,000 miles, not due east, but along a great circle which lands you at a point due east of the Shetland Islands. This, however, only reinforces our conclusion: you will end at a point even farther east of your starting point than before.

What are the differences between the geometry on a sphere and the geometry on a plane? If you make a triangle on the earth, whose sides are great circles, you will not find that the angles of the triangle add up to two right angles:

they will add up to rather more. The amount by which they exceed two right angles is proportional to the size of the triangle. On a small triangle such as you could make with strings on the grass, or even on a triangle formed by three ships which can just see each other, the angles will add up to so little more than two right angles that you will not be able to detect the difference. But if you take the triangle made by the equator, the Greenwich meridian, and the 90th meridian, the angles add up to *three* right angles. And you can get triangles in which the angles add up to anything up to six right angles. All this you could discover by measurements on the surface of the earth, without having to take account of anything in the rest of space.

The theorem of Pythagoras also will fail for distances on a sphere. For the point of view of a traveler bound to the earth, the distance between two places is their great-circle distance, that is to say, the shortest journey that a person can make without leaving the surface of the earth. Now suppose you take three bits of great circles which make a triangle, and suppose one of them is at right angles to another—to be definite, let one be the equator and one a bit of the meridian of Greenwich going northward from the equator. Suppose you go 3,000 miles along the equator and then 4,000 miles due north; how far will you be from your starting point,

Intervals in Space-time

estimating the distance along a great circle? If you were on a plane, your distance would be 5,000 miles, as we saw before. In fact, however, your great-circle distance will be considerably less than this. In a right-angled triangle on a sphere, the square on the side opposite the right angle is less than the sum of the squares on the other two sides.

These differences between the geometry on a sphere and the geometry on a plane are intrinsic differences; that is to say, they enable you to find out whether the surface on which you live is like a plane or like a sphere, without requiring that you should take account of anything outside the surface. Such considerations led to the next step of importance in our subject, which was taken by Gauss, who flourished a hundred and fifty years ago. Gauss studied the theory of surfaces and showed how to develop it by means of measurements on the surfaces themselves, without going outside them. In order to fix the position of a point in space, we need three measurements, but in order to fix the position of a point on a surface we need only two—for example, a point on the earth's surface is fixed when we know its latitude and longitude.

Now Gauss found that, whatever system of measurement you adopt and whatever the nature of the surface, there is always a way of calculating the distance between two not very distant points on the surface, when you know the quantities which fix their positions. The formula for the distance is a generalization of the formula of Pythagoras; it tells you the square of the distance in terms of the squares of the differences between the measure quantities which fix the points, and also the product of these two quantities. When you know this formula, you can discover all the intrinsic properties of the surface, that is to say, all those which do not depend upon its relations to points outside the surface. You can discover, for example, whether the angles of a triangle add up to two right angles, or more, or less, or more in some cases and less in others.

But when we speak of a "triangle," we must explain what we mean, because on most surfaces there are no straight lines. On a sphere, we shall replace straight lines by great circles, which are the nearest possible approach to straight lines. In general, we shall take, instead of straight lines, the lines that give the shortest route on the surface from place to place. Such lines are called "geodesics." On the earth, the geodesics are great circles. In general, they are the shortest way of traveling from point to point if you are unable to leave the surface. They take the place of straight lines in the intrinsic geometry of a surface. When we inquire whether the angles of a triangle add up to two right angles or not, we mean to speak of a triangle whose sides are geodesics. And when we speak of the distance between two points, we mean the distance along a geodesic.

The next step in our generalizing process is rather difficult: it is the transition to non-Euclidean geometry. We live in a world in which space has three dimensions, and our empirical knowledge of space is based upon measurement of small distances and of angles. (When I speak of small distances, I mean distances that are small compared to those in astronomy; all distances on the earth are small in this sense.) It was formerly thought that we could be sure *a priori* that space is Euclidean—for instance, that the angles of a triangle add up to two right angles. But it came to be recognized that we could not prove this by reasoning; if it was to be known, it must be known as the result of measurements. Before Einstein, it was thought that measurements confirm Euclidean geometry within the limits of exactitude attainable; now this is no longer thought. It is still true that we can, by what may be called a natural artifice, cause Euclidean geometry to *seem* true throughout a small region, such as the earth, but in explaining gravitation Einstein is led to the view that over large regions where there is matter we cannot regard space as Euclidean.

The reasons for this will concern us later. What concerns us now is the way in which non-Euclidean geometry results from a generalization of the work of Gauss.

There is no reason why we should not have the same circumstances in three-dimensional space as we have, for example, on the surface of a sphere. It might happen that the angles of a triangle would always add up to more than two right angles, and that the excess would be proportional to the size of the triangle. It might happen that the distance between two points would be given by a formula analogous to what we have on the surface of a sphere, but involving three quantities instead of two. Whether this does happen or not can be discovered only by actual measurements. There are an infinite number of such possibilities.

This line of argument was developed by Riemann, in his dissertation "On the hypotheses which underlie geometry" (1854), which applied Gauss's work on surfaces to different kinds of three-dimensional spaces. He showed that all the essential characteristics of a kind of space could be deduced from the formula for small distances. He assumed that, from the small distances in three given directions which would together carry you from one point to another not far from it, the distances between the two points could be calculated. For instance, if you know that you can get from one point to another by first moving a certain distance east, then a certain distance north, and finally a certain distance straight up in the air, you are to be able to calculate the distance from the one point to the other. And the rule for the calculation is to be an extension of the theorem of Pythagoras, in the sense that you arrive at the square of the required distance by adding together multiples of the squares of the component distances, together possibly with multiples of their products. From certain characteristics in the formula, you can tell what sort of space you have to deal with. These characteristics do not depend upon the particular method you have adopted for determining the positions of points.

In order to arrive at what we want for the theory of relativity, we now have one more generalization to make: we have to substitute the "interval" between events for the distance between points. This takes us to space-time. We have already seen that, in the special theory of relativity, the square of the interval is found by subtracting the square of the distance between events from the square of the distance that light would travel in the time between them. In the general theory, we do not assume this special form of interval. We assume to begin with a general form, like that which Riemann used for distances. Moreover, like Riemann, Einstein only assumes the formula for *neighboring* events, that is to say, events which have only a small interval between them. What goes beyond these initial assumptions depends upon observation of the actual motion of bodies, in ways which we shall explain in later chapters.

We may now sum up and restate the process we have been describing. In three dimensions, the position of a point relatively to a fixed point (the "origin") can be determined by assigning three quantities ("coordinates"). For example, the position of a balloon relatively to your house is fixed if you know that you will reach it by going first a given distance due east, then another given distance due north, then a third given distance straight up. When, as in this case, the three coordinates are three distances all at right angles to each other, which, taken successively, transport you from the origin to the point in question, the square of the direct distance to the point in question is got by adding up the squares of the three coordinates. In all cases, whether in Euclidean or in non-Euclidean spaces, it is got by adding multiples of the squares and products of the coordinates according to an assignable rule. The coordinates may be any quantities which fix the position of a point, provided that neighboring points must have neighboring quantities for their coordinates. In the general theory of relativity, we

Intervals in Space-time

add a fourth coordinate to give the time, and our formula gives "interval" instead of spatial distance; moreover, we assume the accuracy of our formula for small distances only.

We are now at last in a position to tackle Einstein's theory of gravitation.

CHAPTER 8

Einstein's Law of Gravitation

Before tackling Einstein's law, it is as well to convince ourselves, on logical grounds, that Newton's law of gravitation cannot be quite right.

Newton said that between any two particles of matter there is a force which is proportional to the product of their masses and inversely proportional to the square of their distance. That is to say, ignoring for the present the question of mass, if there is a certain attraction when the particles are a mile apart, there will be a quarter as much attraction when they are two miles apart, a ninth as much when they are three miles apart, and so on: the attraction diminishes much faster than the distance increases. Now, of course, Newton, when he spoke of the distance, meant the distance at a given time: he thought there could be no ambiguity about time. But we have seen that this was a mistake. What one observer judges to be the same moment on the earth and the sun, another will judge to be two different moments. "Distance at a given moment" is therefore a subjective conception, which can hardly enter into a cosmic law. Of course, we could make our law unambiguous by saying that we are going to estimate times as they are estimated by Greenwich Observatory. But we can hardly believe that the accidental circumstances of the earth deserve to be taken so seriously. And the estimate of distance, also, will vary for different observers. We cannot therefore allow that Newton's form of the law of gravitation can be quite correct, since it will give different results according to which of many equally legitimate conventions we adopt. This is as absurd as it would be if the question whether one person had murdered another were to depend upon whether they were described by their first names or their surnames. It is obvious that physical laws must be the same whether distances are measured in miles or in kilometers, and we are concerned with what is essentially only an extension of the same principle.

Our measurements are conventional to an

even greater extent than is admitted by the special theory of relativity. Moreover, every measurement is a physical process carried out with physical material; the result is certainly an experimental datum, but may not be susceptible of the simple interpretation which we ordinarily assign to it. We are, therefore, not going to assume to begin with that we know how to measure anything. We assume that there is a certain physical quantity called "interval," which is a relation between two events that are not widely separated; but we do not assume in advance that we know how to measure it, beyond taking it for granted that it is given by some generalization of the theorem of Pythagoras such as we spoke of in the preceding chapter.

We do assume, however, that events have an *order,* and that this order is four-dimensional. We assume, that is to say, that we know what we mean by saying that a certain event is nearer to another than a third, so that before making accurate measurements we can speak of the "neighborhood" of an event; and we assume that, in order to assign the position of an event in space-time, four quantities (coordinates) are necessary—e.g., in our former case of an explosion on an airship, latitude, longitude, altitude, and time. But we assume nothing about the way in which these coordinates are assigned, except that neighboring coordinates are assigned to neighboring events.

The way in which these numbers, called coordinates, are to be assigned is neither wholly arbitrary nor a result of careful measurement—it lies in an intermediate region. While you are making any continuous journey, your coordinates must never alter by sudden jumps. In America one finds that the houses between (say) Fourteenth Street and Fifteenth Street are likely to have numbers between 1400 and 1500, while those between Fifteenth Street and Sixteenth Street have numbers between 1500 and 1600, even if the 1400s were not used up. This would not do for our purposes, because there is a sudden jump when we pass from one block to the next. Or again we might assign the time coordinate in the following way: take the time that elapses between two successive births of people called Smith; an event occurring between the births of the 3000th and the 3001st Smith known to history shall have a coordinate lying between 3000 and 3001; the fractional part of its coordinate shall be the fraction of a year that has elapsed since the birth of the 3000th Smith. (Obviously there could never be as much as a year between two successive additions to the Smith family.) This way of assigning the time coordinate is perfectly definite, but it is not admissible for our purposes, because there will be sudden jumps between events just before the birth of a Smith and events just after, so that in a continuous journey your time coordinate will not change continuously. It is assumed that, independently of measurement, we know what a continuous journey is. And when your position in space-time changes continuously, each of your four coordinates must change continuously. One, two, or three of them may not change at all, but whatever change does occur must be smooth, without sudden jumps. This explains what is *not* allowable in assigning coordinates.

To explain all the changes that are legitimate in your coordinates, suppose you take a large piece of soft india rubber. While it is in an unstretched condition, measure little squares on it, each one-tenth of an inch each way. Put in little tiny pins at the corners of the squares.

We can take as two of the coordinates of one of these pins the number of pins passed in going to the right from a given pin until we come just below the pin in question, and then the number of pins we pass on the way up to this pin. In the figure, let 0 be the pin we start from and P the pin to which we are going to assign coordinates. P is in the fifth column and the third row, so its coordinates in the plane of the india rubber are to be 5 and 3.

Now take the india rubber and stretch it and twist it as much as you like. Let the pins now be

in the shape they have in the second figure. The divisions now no longer represent distances according to our usual notions, but they will still do just as well as coordinates. We may still take P as having the coordinates 5 and 3 in the plane of the india rubber, and we may still regard the india rubber as being in a plane, even if we have twisted it out of what we should ordinarily call a plane. Such continuous distortions do not matter.

To take another illustration: instead of using a steel measuring rod to fix our coordinates, let us use a live eel, which is wriggling all the time. The distance from the tail to the head of the eel is to count as 1 from the point of view of coordinates, whatever shape the creature may be assuming at the moment. The eel is continuous, and its wriggles are continuous, so it may be taken as our unit of distance in assigning coordinates. Beyond the requirement of continuity, the method of assigning coordinates is purely conventional, and therefore a live eel is just as good as a steel rod.

We are apt to think that, for really careful measurements, it is better to use a steel rod than a live eel. This is a mistake, not because the eel tells us what the steel rod was thought to tell, but because the steel rod really tells no more than the eel obviously does. The point is, not that eels are really rigid, but that steel rods really wriggle. To an observer in just one possible state of motion the eel would appear rigid, while the steel rod would seem to wriggle just as the eel does to us. For everybody moving differently both from this observer and ourselves, both the eel and the rod would seem to wriggle. And there is no saying that one observer is right and another wrong. In such matters what is seen does not belong solely to the physical process observed, but also to the standpoint of the observer. Measurements of distances and times do not directly reveal properties of the things measured, but relations of the things to the measurer. What observation can tell us about the physical world is therefore more abstract than we have hitherto believed.

It is important to realize that geometry, as taught in schools since Greek times, ceases to exist as a separate science and becomes merged into physics. The two fundamental notions in elementary geometry were the straight line and the circle. What appears to you as a straight road, whose parts all exist now, may appear to another observer to be like the flight of a rocket, some kind of curve whose parts come into existence successively. The circle depends upon measurement of distance, since it consists of all the points at a given distance from its center. And measurement of distances, as we have seen, is a subjective affair, depending upon the way in which the observer is moving. The failure of the circle to have objective validity was demonstrated by the Michelson-Morley experiment, and is thus, in a sense, the starting point of the whole theory of relativity. Rigid bodies, which we need for measurement, are only rigid for certain observers; for others they will be constantly changing all their dimensions. It is only our obstinately earthbound imagination that makes us suppose a geometry separate from physics to be possible.

That is why we do not trouble to give physical significance to our coordinates from the start. Formerly, the coordinates used in physics were supposed to be carefully measured distances; now we realize that this care at the start is thrown away. It is at a later stage that care is required. Our coordinates now are hardly more than a systematic way of cataloguing events.

But mathematics provides, in the method of tensors, such an immensely powerful technique that we can use coordinates assigned in this apparently careless way just as effectively as if we had applied the whole apparatus of minutely accurate measurement in arriving at them. The advantage of being haphazard at the start is that we avoid making surreptitious physical assumptions, which we can hardly help making if we suppose that our coordinates have initially some particular physical significance.

We need not try to proceed in ignorance of all observed physical phenomena. We know certain things. We know that the old Newtonian physics is very nearly accurate when our coordinates have been chosen in a certain way. We know that the special theory of relativity is still more nearly accurate for suitable coordinates. From such facts we can infer certain things about our new coordinates, which, in a logical deduction, appear as postulates of the new theory.

As such postulates we take:

1 That the interval between neighboring events takes a general form, like that used by Riemann for distances.
2 That a sufficiently small, light, and symmetrical body travels on a geodesic in space-time, except in so far as nongravitational forces act upon it.
3 That a light ray travels on a geodesic which is

such that the interval between any two parts of it is zero.

Each of these postulates requires some explanation.

Our first postulate requires that, if two events are close together (but not necessarily otherwise), there is an interval between them which can be calculated from the differences between their coordinates by some such formula as we considered in the preceding chapter. That is to say, we take the squares and products of the differences of coordinates, we multiply them by suitable amounts (which in general will vary from place to place), and we add the results together. The sum obtained is the square of the interval. We do not assume in advance that we know the amounts by which the squares and products must be multiplied; this is going to be discovered by observing physical phenomena. But we do know, because mathematics shows it to be so, that within any small region of space-time we can choose the coordinates so that the interval has almost exactly the special form which we found in the special theory of relativity. It is not necessary for the application of the special theory to a limited region that there should be no gravitation in the region; it is enough if the intensity of gravitation is practically the same throughout the region. This enables us to apply the special theory within any small region. How small it will have to be depends upon the neighborhood. On the surface of the earth, it would have to be small enough for the curvature of the earth to be negligible. In the spaces between the planets, it need only be small enough for the attraction of the sun and the planets to be sensibly constant throughout the region. In the spaces between the stars it might be enormous—say half the distance from one star to the next—without introducing measurable inaccuracies.

Thus, at a great distance from gravitating matter, we can so choose our coordinates as to obtain very nearly a Euclidean space; this is really only another way of saying that the special theory of relativity applies. In the neighborhood of matter, although we can still make our space very nearly Euclidean in a very small region, we cannot do so throughout any region within which gravitation varies sensibly—at least, if we do, we shall have to abandon the view expressed in the second postulate, that bodies moving under gravitational forces only move on geodesics.

We saw that a geodesic on a surface is the shortest line that can be drawn on the surface from one point to another—for example, on the earth the geodesics are great circles. When we come to space-time, the mathematics is the same, but the verbal explanations have to be rather different. In the general theory of relativity, it is only neighboring events that have a definite interval, independently of the route by which we travel from one to the other. The interval between distant events depends upon the route pursued, and has to be calculated by dividing the route into a number of little bits and adding up the intervals for the various little bits. If the interval is spacelike, a body cannot travel from one event to the other; therefore, when we are considering the way bodies move, we are confined to timelike intervals. The interval between neighboring events when it is timelike will appear as the time between them for observers who travel from the one event to the other. And so the whole interval between two events will be judged by people who travel from one to the other to be what their clocks show to be the time that they have taken on the journey. For some routes this time will be longer, for others shorter; the more slowly they travel, the longer they will think they have been on the journey. This must not be taken as a platitude. I am not saying that if you travel from London to Edinburgh you will take longer if you travel more slowly. I am saying something much more odd. I am saying that if you leave London at 10 A.M. and arrive in Edinburgh at 6.30 P.M., Greenwich time, the more slowly you travel, the longer you will take—if the time is judged by your watch. This is a very different statement. From the point of view of a person on the earth,

your journey takes eight hours and a half. But if
you had been a ray of light traveling around the
solar system, starting from London at 10 A.M.,
reflected from Jupiter to Saturn, and so on, until at last you were reflected back to Edinburgh
and arrived there at 6.30 P.M., you would judge
that the journey had taken you exactly no time.
And if you had gone by any circuitous route,
which enabled you to arrive in time by traveling fast, the longer your route, the less time you
would judge that you had taken; the diminution
of time would be continual as your speed ap-
proached that of light. Now I say that when a
body travels, if it is left to itself, it chooses the
route which makes the time between two stages
of the journey as long as possible; if it had traveled from one event to another by any other
route, the time, as measured by its own clocks,
would have been shorter. This is a way of saying
that bodies left to themselves do their journeys
as slowly as they can; it is a sort of law of cosmic
laziness. Its mathematical expression is that
they travel in geodesics, in which the total interval between any two events on the journey is
greater than by any alternative route. (The fact
that it is greater, not less, is due to the fact that
the sort of interval we are considering is more
analogous to time than to distance.) For example, if people could leave the earth and travel
about for a time and then return, the time between their departure and return would be less

by their clocks than by those on the earth: the earth, in its journey around the sun, chooses the route which makes the time of any bit of its course by its clocks longer than the time as judged by clocks which move by a different route. This is what is meant by saying that bodies left to themselves move in geodesics in space-time.

It is important to remember that space-time is not supposed to be Euclidean. As far as the geodesics are concerned, this has the effect that space-time is like a hilly countryside. In the neighborhood of a piece of matter, there is, as it were, a hill in space-time; this hill grows steeper and steeper as it gets nearer the top, like the neck of a bottle. It ends in a sheer precipice. Now, by the law of cosmic laziness which we mentioned earlier, a body coming into the neighborhood of the hill will not attempt to go straight over the top, but will go around. This is the essence of Einstein's view of gravitation. What a body does, it does because of the nature of space-time in its own neighborhood, not because of some mysterious force emanating from a distant body.

An analogy will serve to make the point clear. Suppose that on a dark night a number of people with lanterns were walking in various directions across a huge plain, and suppose that in one part of the plain there was a hill with a flaring beacon on the top. Our hill is to be such as we have described, growing steeper as it goes up and ending in a precipice. I shall suppose that there are villages dotted about the plain, and the people with lanterns are walking to and from these various villages. Paths have been made showing the easiest way from any one village to any other. These paths will all be more or less curved, to avoid going too far up the hill; they will be more sharply curved when they pass near the top of the hill than when they keep some way off from it. Now, suppose that you are observing all this, as best you can, from a place high up in a balloon, so that you cannot see the ground, but only the lanterns and the beacon. You will not know that there is a hill or that the beacon is at the top of it. You will see that people turn out of the straight course when they approach the beacon, and that the nearer they come, the more they turn aside. You will naturally attribute this to an effect of the beacon; you may think that it is very hot and people are afraid of getting burned. But if you wait for daylight you will see the hill, and you will find that the beacon merely marks the top of the hill and does not influence the people with lanterns in any way.

Now in this analogy the beacon corresponds to the sun, the people with lanterns correspond to the planets and comets, the paths correspond to their orbits, and the coming of daylight corresponds to the coming of Einstein. Einstein says that the sun is at the top of a hill, only the hill is in space-time, not in space. (I advise the reader not to try to picture this, because it is impossible.) Each body, at each moment, adopts the easiest course open to it, but owing to the hill, the easiest course is not a straight line. Each little bit of matter is at the top of its own little hill, like the cock on his own dung heap. What we call a big bit of matter is a bit which is the top of a big hill. The hill is what we know about; the bit of matter at the top is assumed for convenience. Perhaps there is really no need to assume it, and we could do with the hill alone, for we can never get to the top of anyone else's hill, any more than the pugnacious cock can fight the peculiarly irritating bird that he sees in the looking glass.

I have given only a qualitative description of Einstein's law of gravitation; to give its exact quantitative formulation is impossible without more mathematics than I am permitting myself. The most interesting point about it is that it makes the law no longer the result of action at a distance; the sun exerts no force on the planets whatever. Just as geometry has become physics, so, in a sense, physics has become geometry. The law of gravitation has become the geometrical law that every body pursues the easiest course from place to place, but this course is affected by the hills and valleys that are encountered on the road.

We have been assuming that the body considered is acted upon only by gravitational forces. We are concerned at present with the law of gravitation, not with the effects of electromagnetic forces or the forces between subatomic particles. There have been many attempts to bring all these forces into the framework of general relativity, by Einstein himself, and by Weyl, Kaluza, and Klein, to mention only a few of the others, but none of these attempts has been entirely satisfactory. For the present, we may ignore this work, because the planets are not subject, as wholes, to appreciable electromagnetic or subatomic forces; it is only gravitation that has to be considered in accounting for their motions, with which we are concerned in this chapter.

Our third postulate, that a light ray travels so that the interval between two parts of it is zero, has the advantage that it does not have to be stated only for *small* distances. If each little bit of interval is zero, the sum of them all is zero, and so even distant parts of the same light ray have a zero interval. The course of a light ray is also a geodesic according to this postulate. Thus we now have two empirical ways of discovering what are the geodesics in space-time, namely light rays and bodies moving freely. Among freely moving bodies are included all which are not subject, as wholes, to appreciable electromagnetic or subatomic forces—that is to say, the sun, stars, planets, and satellites, and also falling bodies on the earth, at least when they are falling in a vacuum. When you are standing on the earth, you are subject to electromagnetic forces: the electrons and protons in the neighborhood of your feet exert a repulsion on your feet which is just enough to overcome the earth's gravitation. This is what prevents you from falling through the earth, which, solid as it looks, is mostly empty space.

CHAPTER 9

*Proofs of Einstein's Law
of Gravitation*

The reasons for accepting Einstein's law of gravitation rather than Newton's are partly empirical, partly logical. We will begin with the former.

The new law of gravitation gives very nearly the same results as the old when applied to the calculation of the orbits of the planets and their satellites. If it did not, it could not be true, since the consequences deduced from the old law have been found to be almost exactly verified by observation. When, in 1915, Einstein first published the new law, there was only one empirical fact to which he could point to show that his theory was better than the old one. This was what is called the motion of the perihelion of Mercury.

The planet Mercury, like the other planets, moves around the sun in an ellipse, with the sun in one of the foci. At some points of its orbit it is nearer to the sun than at other points. The point where it is nearest to the sun is called its "perihelion." Now, it was found by observation that, from one occasion when Mercury is nearest to the sun until the next, Mercury does not go exactly once around the sun, but a little bit more. The discrepancy is very small; it amounts to an angle of forty-two seconds in a century. Since Mercury goes around the sun rather more than four hundred times a century, it must move about one-tenth of a second of angle more than a complete revolution to get from one perihelion to the next. This very minute discrepancy from Newtonian theory had puzzled astronomers. There was a calculated effect due to perturbations caused by the other planets, but this small discrepancy was the residue after allowing for these perturbations. The new theory accounted exactly for this residue. There is a similar effect in the case of the other planets, but it is much smaller and has not yet been observed with certainty. The perihelion effect in the motion of Mercury was, at first, the new theory's only empirical advantage over the old.

The second success was more sensational. According to orthodox opinion, light in a vacuum ought always to travel in straight lines. Not being composed of material particles, it ought to be unaffected by gravitation. However, it was possible without any serious breach with old ideas to admit that, in passing near the sun, light might be deflected out of the straight path as much as if it were composed of material particles. According to the new theory, however, light would be deflected twice as much as this. That is to say, if the light of a star passed very near the sun, the ray from the star would be turned through an angle of just under one second and three-quarters. Traditionalists were willing to concede half of this amount. Unfortunately, stars which are almost in line with the sun can be seen only during a total eclipse, and even then there may be no sufficiently bright stars near to the sun. Eddington pointed out that, from this point of view, the best day of the year is May 29, because then there are a number of bright stars close to the sun. It happened by incredible good fortune that there was a total eclipse of the sun on May 29, 1919. Two British expeditions photographed the stars near the sun during the eclipse, and the results appeared to confirm the prediction of the new theory. This caused great excitement at the time, but there were many possible sources of error in the observations, and the results cannot be regarded as conclusive. In subsequent eclipse observations the results have varied between one-half and twice the value predicted by the new theory.

However, it has recently been discovered that among the strong starlike sources of radio waves, called quasars, there are some whose emissions pass quite close to the sun, as seen from the earth, at certain times of year. The prediction of the new theory about the deflection of light applies equally to the deflection of radio waves, and by using two or more radio telescopes twenty miles or so apart it is possible to measure the deflection with great precision. The results agree closely with the predictions of the new theory.

The third experimental prediction of the new theory has also been confirmed very precisely, although the experiment is no longer carried out in the way originally proposed by Einstein. Before explaining the effect in question, a few preliminary explanations are necessary. The spectrum of an element consists of certain lines of various shades of light, emitted by the element when it glows, which may be separated by a prism. They are the same (to a close approximation) whether the element is on the earth, the sun, or a star. Each line is of some definite shade of color, with some definite wavelength. Longer wavelengths are toward the red end of the spectrum, shorter ones toward the violet end. When the distance between you and the light source is decreasing, the apparent wavelengths grow shorter, just as waves at sea come quicker when you are traveling ing against the wind. When the distance is increasing, the apparent wavelengths grow longer, for the same reason. This enables us to know whether the stars are becoming nearer to or more distant from us. If the separation is decreasing, all the lines in the spectrum of an element are moved a little toward violet; if increasing, toward red. You may notice the analogous effect in sound any day. If you are in a station and an express comes through whistling, the note of the whistle seems much more shrill while the train is approaching you than when it has passed. Probably many people think the note has "really" changed, but in fact the change in what you hear is only due to the fact that the train was first approaching and then receding. To people in the train, there was no change of note. This is *not* the effect about which the prediction was made. According to the new theory, any periodic process which takes place in an atom lasts for the same time "interval," wherever the atom may be. But a time interval in one place does not correspond exactly to the same time interval somewhere else; this is due to the "hilly" character of space-time which constitutes gravitation.

The theory predicts that a periodic process which takes place in an atom at ground level will take place at a slightly slower rate than it would in a similar atom at the top of a tall building. The emission of light waves is in effect a periodic process; if it takes place more slowly, then it will allow more space between successive wave crests and so produce light of a longer wavelength. Consequently, any given line in the spectrum will, when the light is sent from the ground to the top of a building, seem to observers at the top a little nearer the red end of the spectrum than if the light came from a source at their own level.

Einstein's prediction involved the comparison of light waves emitted by atoms on the sun with light waves emitted by atoms on the earth. The gravitational field at the surface of the sun is very strong compared to that at the surface of the earth, so that the difference in wavelength is greater than between ground level and the top of a building, but the difficulties of measuring the effect in sunlight are so great that the results were inconclusive. The same is the case for measurements of starlight, for which the effect should also occur. At the time of the original prediction, a terrestrial measurement would have been out of the question, but in the last twenty-five years new methods have been invented which make it possible to send light signals whose wavelengths are known with immense precision, and the predicted effect has now been accurately confirmed by many different experiments.

There are many other differences between

the new law of gravitation and the old, some of which have been decisively confirmed by experiment. One of the most precise of these is the time-delay effect, which was not predicted until 1964, almost fifty years after the new theory was proposed.

The reason for this may be that the time delay in question is no more than a few hundred millionths of a second, and the measurement of such short times has only recently become possible. The prediction is that it will take a light signal longer to travel from one chosen place to another if there is a gravitational "hill" nearby than if not. In the experiments, radar signals, to which the prediction applies equally, are sent from the earth to one of the other planets, or to an artificial satellite, and reflected back to earth. The measurements are made when the reflecting agent is on the farther side of the sun, which acts as the gravitational hill. The results confirm the predictions of the theory very exactly, in some cases to within one part in a thousand.

The above experimental tests are quite sufficient to convince astronomers that, where the new theory and the old differ as to the motions of the heavenly bodies, it is the new one that gives the right results. Even if the empirical grounds in favor of the new theory stood alone, they would be conclusive. Whether the new law represents the exact truth or not, it is certainly more nearly exact than the old, though the inaccuracies in the old law were all exceedingly minute.

But the considerations which originally led to the discovery of the new law were not of this detailed kind. Even the consequence about the perihelion of Mercury, which could be verified at once from previous observations, could only be deduced after the theory was complete, and could not form any part of the original grounds for inventing such a theory. These grounds were of a more abstract logical character. I do not mean that they were not based upon observed facts, and I do not mean that they were *a priori* fantasies such as philosophers indulged in formerly. What I mean is that they were derived from certain general characteristics of physical experience, which showed that the old law *must* be wrong and that something like the new law *must* be substituted.

The arguments in favor of the relativity of motion are, as we saw in earlier chapters, quite conclusive. In daily life, when we say that something moves, we mean that it moves relatively to the earth. In dealing with the motions of the planets, we consider them as moving relatively to the sun, or to the center of mass of the solar system. When we say that the solar system itself is moving, we mean that it is moving relatively to the stars. There is no physical occurrence which can be called "absolute motion." Consequently, the laws of physics must be concerned with relative motions, since these are the only kind that occur.

We now take the relativity of motion in conjunction with the experimental fact that the velocity of light is the same relatively to one body as relatively to another, however the two may be moving. This leads us to the relativity of distances and times. This in turn shows that there is no objective physical fact which can be called "the distance between two bodies at a given time," since the time and the distance will both depend on the observer. Therefore, the old law of gravitation is logically untenable, since it makes use of "distance at a given time."

This shows that we cannot rest content with the old law, but it does not show what we are to put in its place. Here several considerations enter in. We have in the first place what is called "the equality of gravitational and inertial mass." What this means is as follows: when you apply a given force* to a heavy body, you do not give it as much acceleration as you would to a light

---

*It has been pointed out already that in the new theory, "gravitational force" is no longer to be regarded as one of the fundamental concepts of dynamics, but only as a convenient way of speaking, it can still be employed like "sunrise" and "sunset," provided we realize what we mean. Often it would require very roundabout expressions to avoid the term "force."

body. What is called the "inertial" mass of a body is measured by the amount of force required to produce a given acceleration. At a given point of the earth's surface, the "mass" is proportional to the "weight." What is measured by scales is rather the mass than the weight: the weight is defined as the force with which the earth attracts the body. Now, this force is greater at the poles than at the equator, because at the equator the rotation of the earth produces a "centrifugal force" which partially counteracts gravitation. The force of the earth's attraction is also greater on the surface of the earth than it is at a great height or at the bottom of a very deep mine. None of these variations is shown by scales, because they affect the weights used just as much as the body weighed, but they are shown if we use a spring balance. The mass does not vary in the course of these changes of weight.

The "gravitational" mass is differently defined. It is capable of two meanings. We may mean (1) the way a body responds in a situation where gravitation has a known intensity, for example, on the surface of the earth, or on the surface of the sun; or (2), the intensity of the gravitational force produced by the body, as, for example, the sun produces stronger gravitational forces than the earth does. The old theory says that the force of gravitation between two bodies is proportional to the product of their masses. Now, let us consider the attraction of different bodies to one and the same body, say the sun. Then different bodies are attracted by forces which are proportional to their masses, and which, therefore, produce exactly the same acceleration in all of them. Thus if we mean "gravitional mass" in sense (1)—that is to say, the way a body responds to gravitation—we find that "the equality of inertial and gravitational mass," which sounds formidable, reduces to this: that in a given gravitational situation, all bodies behave exactly alike. As regards the surface of the earth, this was one of the first discoveries of Galileo. Aristotle thought that heavy bodies fall faster than light ones; Galileo showed that this is not the case, when the resistance of the air is eliminated. In a vacuum, a feather falls as fast as a lump of lead. As regards the planets, it was Newton who established the corresponding facts. At a given distance from the sun, a comet, which has a very small mass, experiences exactly the same acceleration toward the sun as a planet experiences at the same distance. Thus the way in which gravitation affects a body depends only upon where the body is, and in no degree upon the nature of the body. This suggests that the gravitational effect is a characteristic of the locality, which is what Einstein made it.

As for the gravitational mass in sense (2)— i.e., the intensity of the force produced by a body—the new theory predicts that it is the same as the gravitational mass in sense (1). At least one experiment has been done which confirms this prediction.

We have another indication as to what sort of thing the law of gravitation *must* be, if it is to be a characteristic of a neighborhood, as we have seen reason to suppose that it is. It must be expressed in some law which is unchanged when we adopt a different kind of coordinates. We saw that we must not, to begin with, regard our coordinates as having any physical significance: they are merely systematic ways of naming different parts of space-time. Being conventional, they cannot enter into physical laws. That means to say that, if we have expressed a law correctly in terms of one set of coordinates, it must be expressed by the same formula in terms of another set of coordinates. Or, more exactly, it must be possible to find a formula which expresses the law, and which is unchanged however we change the coordinates. It is the business of the theory of tensors to deal with such formulas. And the theory of tensors shows that there is one formula which appears more appropriate than others as being possibly the law of gravitation. When this possibility is examined, it is found to give the right results; it is here that the empirical confirmations come in. But if the new law had not been found to agree with experience, we could not have gone back to the old one. We should have been compelled by logic to seek some law incorporating the relativity of motions, distances, and times, and expressed in terms of "tensors." It is impossible without mathematics to explain the theory of tensors; the nonmathematician must be content to know that it is the technical method by which we eliminate the conventional element from our measurements and laws, and thus arrive at physical laws which are independent of the observer's point of view. Of this method, Einstein's law of gravitation is the most splendid example.

CHAPTER 10

Mass, Momentum,
Energy, and Action

The pursuit of quantitative precision is as arduous as it is important. Physical measurements are made with extraordinary exactitude; if they were made less carefully, such minute discrepancies as form the experimental data for the theory of relativity could never be revealed. Mathematical physics, before the coming of relativity, used a set of conceptions which were supposed to be as precise as physical measurements, but it has turned out that they were logically defective and that this defectiveness showed itself in very small deviations from expectations based upon calculation. In this chapter I want to show how the fundamental ideas of prerelativity physics are affected, and what modifications they have had to undergo.

We have already had occasion to speak of mass. For purposes of daily life, mass is much the same as weight; the usual measures of weight—ounces, grams, etc.—are really measures of mass. But as soon as we begin to make accurate measurements, we are compelled to distinguish between mass and weight. Two different methods of weighing are in common use—one, that of scales; the other, that of the spring balance. When you go a journey and your luggage is weighed, it is not put on scales, but on a spring; the weight depresses the spring a certain amount, and the result is indicated by a needle on a dial. The same principle is used in automatic machines for finding your weight. The spring balance shows weight, but scales show *mass.* So long as you stay in one part of the world, the difference does not matter, but if you test two weighing machines of different kinds in a number of different places, you will find, if they are accurate, that their results do not always agree. Scales will give the same result anywhere, but a spring balance will not. That is to say, if you have a lump of lead weighing ten pounds by the scales, it will also weigh ten pounds by scales in any other part of the world. But if it weighs ten pounds by a spring balance in London, it will weigh more at the North Pole, less at the equator, less high up in an airplane, and less at the bottom of a coal mine, if it is weighed in all those places on the same spring balance. The fact is that the two instruments measure quite different quantities. The scales measure what may be called (apart from refinements which will concern us presently) "quantity of matter." There is the same "quantity of matter" in a pound of feathers as in a pound of lead. Standard "weights," which are really standard "masses," will measure the amount of mass in any substance put into the opposite scales. But "weight" is a property due to the earth's gravitation: it is the amount of the force by which the earth attracts a body. This force varies from place to place. In the first place, anywhere outside the earth the attraction varies inversely as the square of the distance from the center of the earth; it is therefore less at great heights. In the second place, when you go down a coal mine, part of the earth is above you and attracts matter upward instead of downward, so that the net attraction downward is less than on the surface of the earth. In the third place, owing to the rotation of the earth, there is what is called a "centrifugal force," which acts against gravitation. This is greatest at the equator, because there the rotation of the earth involves the fastest motion; at the poles it does not exist, because they are on the axis of rotation. For all these reasons, the force with which a given body is attracted to the earth is measurably different at different places. It is this force that is measured by a spring balance; that is why a spring balance gives different results in different places. In the case of scales, the standard "weights" are altered just as much as the body to be weighed, so that the result is the same everywhere, but the result is the "mass," not the "weight." A standard "weight" has the same mass everywhere, but not the same "weight"; it is in fact a unit of mass, not of weight. For theoretical purposes, mass, which is almost invariable for a given body, is much more important than weight, which varies according to circumstances. Mass may be regarded, to begin with, as "quantity of matter"; we shall see that this view is not strictly correct, but it will serve as a starting point for subsequent refinements.

For theoretical purposes, mass is defined as being determined by the amount of force required to produce a given acceleration: the more massive a body is, the greater will be the force required to alter its velocity by a given amount in a given time. It takes a more powerful engine to make a long train attain a speed of ten miles an hour at the end of the first half-minute than it does to make a short train do so. Or we may have circumstances where the force is the same for a number of different bodies; in that case, if we can measure the accelerations produced in them, we can tell the ratios of their masses: the greater the mass, the smaller the acceleration. We may take, in illustration of this method, an example which is important in connection with relativity. Radioactive bodies emit electrons with enormous velocities. We can observe their path by making them travel through water vapor and form a cloud as they go. We can at the same time subject them to known electric and magnetic forces, and observe how much they are bent out of a straight line by these forces. This makes it possible to compare their masses. It is found that the faster they travel, the greater are their masses as measured by the stationary observer. It is known otherwise that, apart from the effect of motion, all electrons have the same mass.

All this was known before the theory of relativity was invented, but it showed that the traditional conception of mass had not quite the definiteness that had been ascribed to it. Mass used to be regarded as "quantity of matter," and supposed to be quite invariable. Now, mass was found to be relative to the observer, like length and time, and to be altered by motion in exactly the same proportion. However, this could be remedied. We could take the "proper mass," the mass as measured by an observer who shares the motion of the body. This was easily inferred from the measured mass, by taking the same proportion as in the case of lengths and times.

But there is a more curious fact, and that is that after we have made this correction we still have not obtained a quantity which is at all times exactly the same for the same body. When a body absorbs energy—for example, by growing hotter—its "proper mass" increases slightly. The increase is very slight, since it is measured by dividing the increase of energy by the square of the velocity of light. On the other hand, when a body parts with energy, it loses mass. The most notable case of this is that four hydrogen atoms can come together to make one helium atom, but a helium atom has rather less than four times the mass of one hydrogen atom. This phenomenon is of the greatest practical importance. It is thought to occur in the interior of stars, providing the energy which we see as starlight and which, in the case of the sun, supports terrestrial life. It can be also made to occur in terrestrial laboratories, with an enormous liberation of energy in the form of light and heat. This makes possible the manufacture of hydrogen bombs, which are virtually unlimited in size and destructive power. Ordinary atomic bombs, which work by the disintegration of uranium, have a natural limitation: if too much uranium is collected into one place, it is liable to explode by itself without waiting to be detonated, so that uranium bombs cannot be made with more than a certain maximum size. But a hydrogen bomb may contain as much hydrogen as we please, because hydrogen by itself is not explosive. It is only when the hydrogen is detonated by a conventional uranium bomb that it combines to form helium and release energy. This is because the combination can only take place at a very high temperature.

There is a further advantage: the supply of uranium in the planet is limited, and it might be feared that it would be used up before the human race were exterminated, but now that the practically unlimited supply of hydrogen can be utilized, there is considerable reason to hope that the race may put an end to itself, to the great advantage of such less ferocious animals as may survive.

But it is time to return to less cheerful topics.

We have thus two kinds of mass, neither of which quite fulfills the old ideal. The mass as measured by an observer who is in motion relative to the body in question is a relative quantity and has no physical significance as a property of the body. The "proper mass" is a genuine property of the body, not dependent upon the observer; but it, also, is not strictly constant. As we shall see shortly, the notion of mass becomes absorbed into the notion of energy; it represents, so to speak, the energy which the body expends internally, as opposed to that which it displays to the outer world.

Conservation of mass, conservation of momentum, and conservation of energy were the great principles of classical mechanics. Let us next consider conservation of momentum.

The momentum of a body in a given direction is its velocity in that direction multiplied by its mass. Thus a heavy body moving slowly may have the same momentum as a light body moving fast. When a number of bodies interact in any way—for instance, by collisions or by mutual gravitation—so long as no outside influences come in, the total momentum of all the bodies in any direction remains unchanged. This law remains true in the theory of relativity. For different observers, the mass will be different, but so will the velocity; these two differences neutralize each other, and it turns out that the principle still remains true.

The momentum of a body is different in different directions. The ordinary way of measuring it is to take the velocity in a given direction (as measured by the observer) and multiply it by the mass (as measured by the observer). Now the velocity in a given direction is the distance traveled in that direction in unit time. Suppose we take instead the distance traveled in that direction while the body moves through unit "interval." (In ordinary cases, this is only a very slight change, because, for velocities considerably less than that of light, interval is very nearly equal to lapse of time.) And suppose that instead of the mass as measured by the observer we take the proper mass. These two changes increase the velocity and diminish the mass, both in the same proportion. Thus the momentum remains the same, but the quantities that vary according to the observer have been replaced by quantities which are fixed independently of the observer—with the exception of the distance traveled by the body in the given direction.

When we substitute space-time for time, we find that the measured mass (as opposed to the proper mass) is a quantity of the same kind as the momentum in a given direction; it might be called the momentum in the time-direction. The measured mass is obtained by multiplying the invariant mass by the *time* traversed in traveling through unit interval; the momentum is obtained by multiplying the same invariant mass by the *distance* traversed (in the given direction) in traveling through unit interval. From a space-time point of view, these naturally belong together.

Although the measured mass of a body depends upon the way the observer is moving relatively to the body, it is nonetheless a very important quantity. The conservation of measured mass is the same thing as the conservation of energy. This may seem surprising, since at first sight mass and energy are very different things. But it has turned out that energy is the same thing as measured mass. To explain how this comes about is not easy: nevertheless, we will make the attempt.

In popular talk, "mass" and "energy" do not mean at all the same thing. We associate "mass" with the idea of a fat person in a chair, very slow to move, while "energy" suggests a thin person full of hustle and pep. Popular talk associates "mass" with "inertia," but its view of inertia is one-sided: it includes slowness in beginning to move, but not slowness in stopping, which is equally involved. All these terms have technical meanings in physics, which are only more or less analogous to the meanings of the terms in popular talk. For the present, we are concerned with the technical meaning of "energy."

Throughout the latter half of the nineteenth century, a great deal was made of the "conservation of energy," or the "persistence of force," as Herbert Spencer preferred to call it. This principle was not easy to state in a simple way, because of the different forms of energy, but the essential point was that energy is never created or destroyed, though it can be transformed from one kind into another. The principle acquired its position through Joule's discovery of the "mechanical equivalent of heat," which showed that there was a constant proportion between the work required to produce a given amount of heat and the work required to raise a given weight through a given height: in fact, the same sort of work could be utilized for either purpose according to the mechanism. When heat was

found to consist of motion in molecules, it was
seen to be natural that it should be analogous to
other forms of energy. Broadly speaking, by the
help of a certain amount of theory, all forms of
energy were reduced to two, which were called
respectively "kinetic" and "potential." These
were defined as follows:

The kinetic energy of a particle is half the
mass multiplied by the square of the velocity.
The kinetic energy of a number of particles is
the sum of the kinetic energies of the separate
particles.

The potential energy is more difficult to define. It represents any state of strain, which can
only be preserved by the application of force. To
take the easiest case: if a weight is lifted to a
height and kept suspended, it has potential energy, because, if left to itself, it will fall. Its potential energy is equal to the kinetic energy
which it would acquire in falling through the
same distance through which it was lifted. Similarly, when a comet goes around the sun in a
very eccentric orbit, it moves much faster when
it is near the sun than when it is far from it, so
that its kinetic energy is much greater when it
is near the sun. On the other hand, its potential
energy is greatest when it is farthest from the
sun, because it is then like the stone which has
been lifted to a height. The sum of the kinetic
and potential energies of the comet are constant, unless it suffers collisions or loses some of

its material. We can determine accurately the *change* of potential energy in passing from one position to another, but the total amount of it is to a certain extent arbitrary, since we can fix the zero level where we like. For example, the potential energy of our stone may be taken to be the kinetic energy it would acquire in falling to the surface of the earth, or what it would acquire in falling down a well to the center of the earth, or any assigned lesser distance. It does not matter which we take, so long as we stick to our decision. We are concerned with a profit-and-loss account, which is unaffected by the amount of the assets with which we start.

Both the kinetic and the potential energies of a given set of bodies will be different for different observers. In classical dynamics, the kinetic energy differed according to the state of motion of the observer, but only by a constant amount; the potential energy did not differ at all. Consequently, for each observer the total energy was constant—assuming always that the observers concerned were moving in straight lines with uniform velocities, or, if not, were able to refer their motions to bodies which were so moving. But in relativity dynamics the matter becomes more complicated. The Newtonian ideas of kinetic and potential energy can without much difficulty be adapted to the special theory of relativity. But we cannot profitably adapt the idea of potential energy to the general theory of relativity, nor can we generalize, the idea of kinetic energy, except in the case of a single body. Therefore, the conservation of energy, in the usual Newtonian sense, cannot be maintained. The reason is that the kinetic and potential energies of a system of bodies are inherently ideas which refer to extended regions of space-time. The very wide latitude in choice of coordinates, and the hilly character of space-time, which were explained in Chapter 8, combine to make it very awkward to introduce ideas of this sort into the general theory. There is a conservation law in the general theory, but it is not as useful as the conservation laws in Newtonian mechanics and in the special theory, because it depends on the choice of coordinates in a way which is difficult to understand. We have seen that independence of the choice of coordinates is a guiding principle in the general theory of relativity, and the conservation law is suspect because it conflicts with this principle. Whether this means that conservation is of lesser fundamental importance than was thought hitherto, or whether a satisfactory conservation law still lies hidden in the mathematical complexities of the theory, is a question which has still to be resolved. In the meantime, we must in the general theory be satisfied with the idea of kinetic energy for a single particle only. That is all we shall need in the argument which follows. It should be remembered that these difficulties about the conservation of energy arise only in the general theory, and not in the special theory. Whenever gravitation may be neglected and the special theory becomes applicable, the conservation of energy can be maintained.

What is meant by "conservation" in practice is not exactly what it means in theory. In theory we say that a quantity is conserved when the amount of it in the world is the same at any one time as at any other. But in practice we cannot survey the whole world, so we have to mean something more manageable. We mean that, taking any given region, if the amount of the quantity in the region has changed, it is because some of the quantity has passed across the boundary of the region. If there were no births and deaths, population would be conserved; in that case, the population of a country could only change by emigration or immigration, that is to say, by passing across the boundaries. We might be unable to take an accurate census of China or Central Africa, and therefore we might not be able to ascertain the total population of the world. But we should be justified in assuming it to be constant if, wherever statistics were possible, the population never changed except through people crossing the frontiers. In fact, of course, population is not conserved. A physiologist of my acquaintance once put four mice into a thermos. Some hours later, when he went to take them out, there were eleven of them. But mass is not subject to these fluctuations: the mass of the eleven mice at the end of the time was no greater than the mass of the four at the beginning.

This brings us back to the problem for the sake of which we have been discussing energy. We stated that, in relativity theory, measured mass and energy are regarded as the same thing, and we undertook to explain why. It is now time to embark upon this explanation. But here, as at the end of Chapter 6, the totally unmathematical reader will do well to skip, and begin at the following paragraph.

Let us take the velocity of light as the unit of velocity; this is always convenient in relativity theory. Let m be the proper mass of a particle, v its velocity relative to the observer. Then its measured mass will be

$$\frac{m}{\sqrt{1 - v^2}}$$

while its kinetic energy, according to the usual formula, will be

$$\tfrac{1}{2} m v^2$$

As we saw before, energy only occurs in a profit-and-loss account, so that we can add any constant quantity to it that we like. We may therefore take the energy to be

$$m + \tfrac{1}{2}\, m\, v^2$$

Now if v is a small fraction of the velocity of light, $m + \frac{1}{2}\, mv^2$ is almost exactly equal to $\dfrac{m}{\sqrt{1 - v^2}}$. Consequently, for velocities such as large bodies have, the energy and the measured mass turn out to be indistinguishable within the limits of accuracy attainable. In fact, it is better to alter our definition of energy, and take it to be $\dfrac{m}{\sqrt{1 - v^2}}$, because this is the quantity for which the law analogous to conservation holds. And when the velocity is very great, it gives a better measure of energy than the traditional formula. The traditional formula must therefore be regarded as an approximation, of which the new formula gives the exact version. In this way, energy and measured mass become identified.

I come now to the notion of "action," which is less familiar to the general public than energy, but has become more important in relativity physics, as well as in the quantum theory. (The quantum is a small amount of action.) The word "action" is used to denote energy multiplied by time. That is to say, if there is one unit of energy in a system, it will exert one unit of action in a second, 100 units of action in 100 seconds, and so on; a system which has 100 units of energy will exert 100 units of action in a second, and 10,000 in 100 seconds, and so on. "Action" is thus, in a loose sense, a measure of how much has been accomplished: it is increased both by displaying more energy and by working for a longer time. Since energy is the same thing as measured mass, we may also take action to be measured mass multiplied by time. In classical mechanics, the "density" of matter in any region is the mass divided by the volume—that is to say, if you know the density in a small region, you discover the total amount of matter by multiplying the density by the volume of the small region. In relativity mechanics, we always want to substitute space-time for space; therefore, a "region" must no longer be taken to be merely a volume, but a volume lasting for a time; a small region will be a small volume lasting for a small time. It follows that, given the density, a small region in the new sense contains, not a small mass merely, but a small mass multiplied by a small time—that is to say, a small amount of "action." This explains why it is to be expected that "action" will prove of fundamental importance in relativity mechanics. And so in fact it is.

The postulate that a freely moving particle follows a geodesic may be replaced by an equivalent assumption about the "action" of the particle. Such an assumption is called a *Princi-*

ple of Least Action. This states that, in passing from one state to another, a body chooses a route involving less action than any slightly different route—again a law of cosmic laziness! Principles of least action are not restricted to single bodies. It is possible to make a similar assumption which leads to a description of space-time as a whole, complete with hills and valleys. Such principles, which play a central part in quantum theory as well as in relativity, are the most comprehensive means of stating the purely formal part of mechanics.

CHAPTER 11

The Expanding Universe

We have been dealing hitherto with experiments and observations most of which concern the earth or the solar system. Only occasionally have we had to reach so far afield as the stars. In this chapter we shall range much farther: we shall see what relativity theory has to say about the universe as a whole.

The astronomical observations which we shall be discussing are believed to be in broad terms well-established, but are continually being revised as the introduction of new techniques makes new observations possible. Moreover, the theoretical explanations of these results are quite speculative in character, and it must not be supposed that we are dealing with theoretical matters having the same solidity as those with which we have been concerned hitherto. They certainly need improvement. Science does not aim at establishing immutable truths and eternal dogmas: its aim is to approach the truth by successive approximations, without claiming that at any stage final and complete accuracy has been achieved.

A few preliminary explanations about the general appearance of the universe are necessary. Much is now known about the distribution of matter on a very large scale. Our sun is one star in a system of many millions of stars called the Galaxy. The Galaxy is shaped like a giant Catherine wheel, with spiral arms of stars coming out of a bright central hub. The sun lies in one of the spiral arms of stars, probably about 28,000 light-years out from the center of the hub of the Galaxy. This distance, like most measurements on the galactic scale, is difficult to estimate, and may be revised in the future. (On the galactic scale, distances are often measured in light-years. A light-year is the distance which light travels in a year: you may calculate it by multiplying the speed of light, which is given on p.24, by the number of seconds in a year, which is 31,536,000. If you convert the answer into miles, it is nearly six million million miles.)

The Milky Way, a bright band of stars across

the sky which is easily visible on a clear night, is just our edge-on view of the rest of the Galaxy from our position in the spiral arm.

The outlines of the Galaxy are not at all sharp. The main body of stars is about 120,000 light-years across, but besides stars the Galaxy contains a great deal of gas, mostly hydrogen, and dust, as well as other material which has not yet been identified. This unknown material is believed to form a spherical cloud which extends well beyond the visible distribution of stars. The cloud is not observed directly; its existence is inferred from its gravitational effects on the stars and other observable matter. The cloud may be as much as 500 or 600 thousand light-years across.

The whole accumulation of stars, gas, dust, and unidentified material rotates slowly around the hub. It is from the manner of rotation that the presence of the unidentified material is inferred. The speed of rotation seems to vary with distance from the center of the hub in a way which cannot be explained without this material.

The sun moves relative to the hub with a speed estimated to be near 220 kilometers a second. If the estimates of speed and of distance to the center are correct, then it takes the sun about 240 million years to go once all the way around the hub.

The mass of the galactic material within the

surrounding cloud is thought to be about a million million times the mass of the sun.

The Galaxy is known to have a retinue of satellites, of which the nearest and best-known are the Magellanic Clouds, which lie about two hundred thousand light-years from the center.

The Galaxy is by no means alone in the universe. It is one among many millions of similar systems scattered throughout the region which our telescopes can explore. The other systems are also called galaxies (or sometimes "nebulas"). Some galaxies are flattened, with spiral arms like our own, others are round like basketballs or oval like footballs, still others quite irregular in shape.

Galaxies show a distinct tendency to be collected into groups. These groups are called "clusters." A single cluster may contain hundreds or thousands of individual galaxies, each of which, like our own, may contain many millions of stars. Our Galaxy belongs to a small cluster called the Local Group. The number of members of the local group is not certain, because several possible members are very faint, but it is thought that there are no more than twenty of them. The best-known galaxy in the Local Group, and the nearest spiral, is the Andromeda Galaxy, named after the constellation in which it appears. It is about two million light-years away, and is faintly visible to the naked eye. The Local Group may be about three million light-years across.

Clusters of galaxies are found, in their turn, to be grouped into larger entities, called superclusters. A supercluster may be 30 million to 100 million or more light-years across, and its mass may be as much as ten thousand times the mass of the whole Galaxy. It is at present believed that superclusters are the largest identifiable aggregations of material in the universe. Although it is thought that single clusters are held together by the gravitational attraction between their component galaxies, it is not yet clear whether the same is true of superclusters or no. The existence of the latter is fairly well established by observation, but nothing is known about their development.

Recent observations suggest that between the superclusters there are voids that contain little or no visible material and are probably even larger than the superclusters themselves.

Despite the aggregation of stars into galaxies, galaxies into clusters, and clusters into superclusters, it is usually supposed that on a sufficiently large scale, the universe is approximately uniform and that the part of it which is observable with existing instruments may be typical of the universe as a whole.

This idea that the universe is uniform on a large scale, which was suggested long before there was adequate astronomical evidence for it, has now acquired the status of a fundamental postulate. It is usually called the "cosmological principle." The cosmological principle is really only an extension of Copernicus' ideas. As soon as we give up the egotistical notion that the earth is at the center of all things, we are forced to realize that the sun, which is an ordinary star, has no more claim than the earth to a special place in our description of the universe. When we find that our Galaxy and the cluster to which it belongs are also typical specimens, then they too must be placed logically on a par with other similar objects. Nor is there any empirical reason for supposing that the laws of physics vary systematically from one cluster of galaxies to the next.

The implications of this may be put in a slightly different way. Suppose you were put into a box without windows and transported to a distant part of the universe. When released from the box you would not, of course, see the particular distribution of stars and galaxies which is visible from the earth—the geographical details of your new environment would be different—but according to the cosmological principle, the overall appearance of the universe would be the same. Aside from details, you would not be able to tell what part of the universe you were in.

There is one very remarkable phenomenon which might have led us to suppose that our local cluster of galaxies does after all have a special position in the universe. This is the so-called "redshift" in the spectra of distant galaxies. As we shall now see, it is because of this phenomenon that the universe is said to be expanding.

We are concerned here with an effect which was explained in Chapter 9,* although in that chapter we were not directly concerned with it. You will remember the analogy with sound which was introduced then: if a train is moving toward you, then the pitch of its whistle is higher than if it is standing still, while if it is moving away from you, the pitch is lower. The effects are very similar in the case of light. If the source of light is moving toward you, then the whole spectrum of the light is shifted toward the violet; if the source is moving away from you, then the whole spectrum is shifted toward the red. These shifts of the spectrum correspond to the changes of pitch of the train whistle. The amount of the shift depends on the rate of change of the distance between you and the light source. (This has nothing to do with the speed of the light itself, which as we have seen is independent of the motion of its source.) This shift of the spectrum provides a means of determining the speeds of stars and galaxies, by comparing the spectra of the light

*On pages 131–33.

which they send out with similar spectra produced in laboratories on earth. The speeds of galaxies in the Local Group, measured in this way, range up to about 300 miles a second. This is very fast by everyday standards, but because of the great distances between the galaxies it would be millions of years before there was any noticeable change in their positions.

Some of the galaxies in the Local Group are moving toward us, others away from us. There is nothing very remarkable about this motion, which might be compared to the motion of bees in a swarm. The bees move about relative to one another, but the swarm as a whole keeps together. The situation is rather different when we come to examine clusters other than our own. Here again there are internal motions in each cluster, but all the other clusters appear to be moving *away* from our own, and the farther away they are, the faster they appear to be moving. It is this remarkable phenomenon which suggests that the universe is expanding.

Because all the other clusters appear to be moving away from ours, we might be inclined to think that the Local Group is in some way at the center of the expanding universe. This would be a mistake, because it ignores the relative character of motion which has been pointed out repeatedly in this book. Consider again the analogy with swarms of bees. Suppose that they are very well-trained swarms, which hover above the ground ten yards apart in a line running from west to east. Then suppose that one of the swarms stays at rest relative to the ground, while the swarm ten yards to the east of it moves east at a yard a minute, the swarm twenty yards to the east moves east at two yards a minute, and so on, while the swarms to the west of the fixed swarm move west at similar speeds. Then it will appear to a bee in any of the swarms, fixed or moving, that all the other swarms are receding at speeds proportional to their distances. If the ground were not available as a standard of rest, then there would be no reason to think that any one of the swarms was picked out in a special way.

The behavior of the clusters of galaxies is entirely similar. Of course they are distributed irregularly in all directions instead of being lined up like our well-trained swarms, but as in the case of the swarms, it appears to an observer in any cluster that all the others are receding. Since there is no absolute standard of rest in the universe, the appearance of expansion is the same for all the clusters.

The most distant cluster so far investigated has a redshift corresponding to a speed of recession which is about half the speed of light. (Speeds of recession corresponding to redshifts as large as this must be calculated on the basis of the Lorentz transformation formulas given in Chapter 6.)

The largest astronomical redshifts discovered hitherto are not those of distant clusters, but of the so-called "quasistellar objects" (quasars) whose redshifts correspond to speeds of recession which are more than nine-tenths of the speed of light. However, the nature of these objects is not yet understood, and so they cannot yet be taken properly into account when the astronomical data are used to construct a theoretical model.

Let us now examine how this information about the universe can be fitted into the general relativity theory. We have seen that the gravitational effects of the sun may be described as those of a hill in space-time. A galaxy, a cluster or a supercluster may be represented in the same way, but by a much larger hill, because of its much greater mass. If we tried to incorporate into this description details of the distribution of stars in each galaxy and galaxies in each cluster, we should have a very complicated hill with many peaks and valleys. We could then try to describe the whole universe in a way which could be represented by a space-time with hills, representing the clusters, scattered about in it. Such a description would be mathematically very complicated, because it would include many "geographical" details not essential to a description of the overall appearance of the universe. In order to simplify the description, we begin by constructing models which preserve what seem to be the essential features while leaving out the geographical details. The features which we preserve are the large-scale uniformity and the expansion. The details left out are the precise positions, sizes, and compositions of the individual galaxies, clusters, and superclusters.

Thus we construct model space-times to represent the universe by supposing it to be exactly, instead of approximately, uniform. In these simplified models we imagine matter to be smoothed out into continuous distribution instead of being collected into aggregates with large spaces in between them.

Just as the accumulation of matter into an aggregate can be described by saying that there is a large hill in space-time where we see the aggregate, or by saying that space-time is curved nearby to that aggregate, so the uniform distribution of matter in a smoothed-out model of the universe can be described by saying that space-time is curved uniformly. The effect of smoothing out the matter composing the different clusters is to smooth out the corresponding curvature to produce a slight overall curvature. This overall curvature of the universe is somewhat analogous to the curvature of a sphere in ordinary space, but it is inappropriate to push the analogy of curvature with space-time hills any further, by comparing the overall curvature of space-time with the curvature of the earth, because this can easily become misleading.

The relativity law of gravitation, combined with the smoothing-out assumption—the assumption of exact uniformity—allows us to construct a variety of models of the universe in which the overall curvature takes a variety of different forms. The main effect of this overall curvature is that it implies, in some of the models, that the spectra of distant objects will be shifted toward the red. It is largely a matter of taste whether this redshift is attributed to a recessional motion, or to space-time curvature. The effect will appear in one guise or the other, depending on the coordinate system which is used to describe the universe. What relativity predicts does not, of course, depend on the choice of coordinate system.

The model universes which we have been considering agree more or less well with observations of the overall properties of our own universe. There are others, equally consistent with the new law and with the assumption of uniformity, in which there is a blueshift, corresponding to a contraction of the universe, instead of a redshift. The existence of such models is no reason for rejecting the new theory. It implies that the theory is not complete—some additional assumption is required which will exclude the unwanted models. Various assumptions have been suggested, but so far an entirely satisfactory one has not been found.

Let us examine the consequences of expansion a little further, remembering always that what we say may always be rephrased in terms of space-time curvature if that becomes necessary. The most obvious consequence is that if the universe is, so to say, thinning out—if the clusters of galaxies are getting farther and farther apart, then in the past they must have been closer together than they are now. Suppose we were to take a movie of the expanding universe, over a period of many millions of years, so as to record the whole history of the expansion. If such a film were to be shown backward, then it would show the history of the universe in reverse. Instead of moving away from one another, all the clusters of galaxies would appear to be moving toward one another. As the film ran back, they would get closer and closer together, until presumably they were so close together that there were no gaps between them anymore. Still farther back, we may suppose, even the spaces between the stars would be closed up, all the available space being filled up with highly condensed hot gas out of which the stars could have evolved.

Recent astronomical observations of short radio waves confirm the existence of this highly condensed state. It seems that a certain proportion of the radio energy arriving at receivers on the earth cannot be attributed to emission by stars or by the interstellar gas, but agrees reasonably well with what might be expected to be visible now of the radiation present in the universe at an early stage when all matter was in a highly condensed state.

However, the predictions of theoretical models about this condensed state cannot be trusted too far. What is known of the quantum properties of matter suggests that at a sufficiently early time, these properties would have had important effects. There is no general agreement about exactly when this would have been so, but in any case, relativity theory alone is unable to describe such effects. At present a great deal of effort is being expended on developments of relativity theory and the quantum theory in order to supply a satisfactory description, but it is not yet clear whether any of these developments will be of any lasting significance.

All this is rather speculative; it is very likely that the universe evolved from a highly condensed state, and it is even more likely that such a highly condensed state represents the earliest time about which there will ever be any scientific information. Whether such a state actually occurred or not is not at present under dispute. Unfortunately, some people are inclined to refer to the highly condensed state as "the beginning of the universe" or "the time when the universe was created" or something of that kind. These phrases mean no more than "the earliest time about which there will ever be any scientific information," and it is better to avoid them, because they carry undesirable metaphysical implications.

As things stand at present, certain of the model universes derived from the relativity theory and predicting expansion from a highly condensed state are readily reconciled with the astronomical data. All of them have defects, of which the most obvious is that they give only a smoothed-out picture which does not account for the size or composition of the galaxies and clusters.

The construction of an entirely satisfactory model depends on the resolving of some serious mathematical difficulties; which of the available models is to be preferred at any particular time must depend on the astronomical data.

CHAPTER 12

*Conventions and
Natural Laws*

One of the most difficult matters in all controversy is to distinguish disputes about words from disputes about facts: it ought not to be difficult, but in practice it is. This is quite as true in physics as in other subjects. In the seventeenth century there was a terrific debate as to what "force" is; to us now, it was obviously a debate as to how the word "force" should be defined, but at the time it was thought to be much more. One of the purposes of the method of tensors, which is employed in the mathematics of relativity, is to eliminate what is purely verbal (in an extended sense) in physical laws. It is of course obvious that what depends on the choice of coordinates is "verbal" in the sense concerned. A person punting walks along the boat, but keeps a constant position with reference to the riverbed so long as she or he does not pick up the pole. The Lilliputians might debate endlessly whether the punter is walking or standing still; the debate would be as to words, not as to facts. If we choose coordinates fixed relatively to the boat, the punter is walking; if we choose coordinates fixed relatively to the riverbed, the punter is standing still. We want to express physical laws in such a way that it shall be obvious when we are expressing the same law by reference to two different systems of coordinates, so that we shall not be misled into supposing we have different laws when we only have one law in different words. This may be accomplished by the method of tensors. Some laws which seem plausible in one language cannot be translated into another; these are impossible as laws of nature. The laws that can be translated into *any* coordinate language have certain characteristics: this is a substantial help in looking for such laws of nature as the theory of relativity can admit to be possible. Of the possible laws, we choose the simplest one which predicts the actual motion of bodies correctly; logic and experience combine in equal proportions in obtaining this expression.

But the problem of arriving at genuine laws of nature is not to be solved by the method of tensors alone; a good deal of careful thought is wanted in addition. Some of this has been done; much remains to be done.

To take a simple illustration: suppose, as in the hypothesis of the Lorentz contraction, that lengths in one direction were shorter than in another. Let us assume that a foot rule pointing north is only half as long as the same foot rule pointing east, and that this is equally true of all other bodies. Does such a hypothesis have any meaning? If you have a fishing rod fifteen feet long when it is pointing west, and you then turn it to the north, it will still measure fifteen feet, because your foot rule will have shrunk too. It won't "look" any shorter, because your eye will have been affected in the same way. If you are to find out the change, it cannot be by ordinary measurement: it must be by some such method as the Michelson-Morley experiment, in which the velocity of light is used to measure lengths. Then you still have to decide whether it is simpler to suppose a change of length or a change in the velocity of light. The experimental fact would be that light takes longer to traverse what your foot rule declares to be a given distance in one direction than in another—or, as in the Michelson-Morley experiment, that it ought to take longer but doesn't. You can adjust your measures to such a fact in various ways; in any way you choose to adopt, there will be an element of convention. This element of convention survives in the laws that you arrive at after you have made your decision as to measures, and often it takes subtle and elusive forms. To eliminate the element of convention is, in fact, extraordinarily difficult; the more the subject is studied, the greater the difficulty is seen to be.

A more important example is the question of the size of the electron. We find experimentally that all electrons are the same size. How far is this a genuine fact ascertained by experiment, and how far is it a result of our conventions of measurement? We have here two different comparisons to make: (1) in regard to one electron at different times; (2) in regard to two electrons at the same time. We can then arrive at the comparison of two electrons at different times, by combining (1) and (2). We may dismiss any hypothesis which would affect all electrons equally; for example, it would be useless to suppose that in one region of space-time they were all larger than in another. Such a change would affect our measuring appliances just as much as the things measured, and would therefore produce no discoverable phenomena. This is as much as to say that it would be no change at all. But the fact that two electrons have the same mass, for instance, cannot be regarded as purely conventional. Given sufficient minuteness and accuracy, we could compare the effects of two different electrons upon a third; if they were equal under like circumstances, we should be able to infer equality in a not purely conventional sense.

Eddington described the process concerned in the more advanced portions of the theory of relativity as "world-building." The structure to be built is the physical world as we know it; the economical architect tries to construct it with the smallest possible amount of material. This is a question for logic and mathematics. The greater our technical skill in these two subjects, the more real building we shall do and the less we shall be content with mere heaps of stones. But before we can use in our building the stones that nature provides, we have to hew them into the right shapes: this is all part of the process of building. In order that this may be possible, the raw material must have *some* structure (which we may conceive as analogous to the grain in timber), but almost any structure will do. By successive mathematical refinements, we whittle away our initial requirements until they amount to very little. Given this necessary minimum of structure in the raw material, we find that we can construct from it a mathematical expression which will have the properties that are needed for describing the world we perceive—in particular, the properties of conservation which are characteristic of momentum and energy (or mass). Our raw material consisted merely of events, but when we find that we can build out of it something which, as measured, will seem to be never created or destroyed, it is not surprising that we should come to believe in "bodies." These are really mere mathematical constructions out of events, but owing to their permanence they are practically important, and our senses (which were presumably developed by biological needs) are adapted for noticing them, rather than the crude continuum of events which is theoretically more fundamental. From this point of view, it is astonishing how little of the real world is revealed by physical science: our knowledge is limited, not only by the conventional element, but also by the selectiveness of our perceptual apparatus.

The limitations of knowledge introduced by the selectiveness of our perceptual apparatus may be illustrated by the indestructibility of energy. This has been gradually discovered by experiment, and seemed a well-founded empirical law of nature. Now it turns out that, from our original space-time continuum, we can construct a mathematical expression which will have properties causing it to appear indestructible. The statement that energy is indestructible then ceases to be a proposition of physics and becomes instead a proposition of linguistics and psychology. As a proposition of linguistics: "energy" is the name of the mathematical expression in question. As a proposition of psychology: our senses are such that we notice what is roughly the mathematical expression in question, and we are led nearer and nearer to it as we refine upon our crude perceptions by scientific observation. This is much less than physicists used to think they knew about energy.

The reader may say, What, then, is left of physics? What do we really know about the world of matter? Here we may distinguish three departments of physics. There is first what is included within the theory of relativity, generalized as widely as possible. Next, there are laws which cannot be brought within the scope of relativity. Third, there is what may be called geography. Let us consider each of these in turn.

The theory of relativity, apart from convention, tells us that the events in the universe have a four-dimensional order, and that, between any two events which are near together in this order, there is a relation called "interval," which is capable of being measured if suitable precautions are taken. It tells us also that "absolute motion," "absolute space," and "absolute time" cannot have any physical significance; laws of physics involving these concepts are not acceptable. This is hardly a physical law in itself, but rather a useful rule to enable us to reject some proposed physical laws as unsatisfactory.

Beyond this, there is little in the theory of relativity that can be regarded as physical laws. There is a great deal of mathematics, showing that certain mathematically constructed quan-

Conventions and Natural Laws

tities must behave like the things we perceive, and there is a suggestion of a bridge between psychology and physics in the theory that these mathematically constructed quantities are what our senses are adapted for perceiving. But neither of these things is physics in the strict sense.

The part of physics which cannot, at present, be brought within the scope of relativity is large and important. There is nothing in relativity to show why there should be electrons and protons; relativity cannot give any reason why matter should exist in little lumps. This is the province of the quantum theory, which accounts for many of the properties of matter on the small scale. The quantum theory has been made consistent with the special theory of relativity, but hitherto all attempts to perform a synthesis of quantum theory and general relativity have been unsuccessful. There seem to be very severe difficulties in the way of bringing this part of physics within the framework of general relativity. At present there are equally severe difficulties in the quantum theory itself, and many physicists think that a synthesis of quantum theory and general relativity might solve some of these difficulties. The present situation, as we have seen, is that general relativity accounts fairly satisfactorily for the properties of matter on a very large scale, while quantum theory accounts fairly satisfactorily for the properties of matter on a very small scale. There is some common ground between the two theories, but the work which has been done to unify them is still to be regarded as speculative. A few people think that general relativity could be extended in such a way as to explain all the results that quantum theory explains, but in a more satisfactory way than present quantum theory does. Einstein toward the end of his life was one of the people who thought this. However, most physicists nowadays think that view is mistaken.

General relativity is the most extreme example of what may be called next-to-next methods. Gravitation need no longer be regarded as due to the effect of the sun on a planet, but may be thought of as expressing the characteristics of the region in which the planet happens to be. These characteristics are supposed to alter bit by bit, gradually, continuously, and not by sudden jumps, as one moves from one part of space-time to another. The effects of electromagnetism may be regarded in a similar way, but as soon as electromagnetism is made to accord with the quantum theory, its character changes entirely. The continuous aspect disappears and is replaced by the discontinuous behavior which as we have already seen is typical of quantum theory. However, if we try to apply to gravitation these ideas of quantum theory we find that they do not fit properly, and that some considerable alteration in one theory or the other, or both, is necessary. What modification is needed is not yet known.

The difficulty may be explained in a somewhat different way. When an astronomer observes the sun, the sun preserves a lordly indifference to the observation. But when a physicist tries to find out what is happening to an atom, the apparatus which is used is much larger than the thing which is observed, instead of much smaller, and is likely to have some effect on it. It is found that the sort of apparatus best suited for determining the position of an atom is bound to affect its velocity, and the sort of apparatus best suited for determining the velocity is bound to affect its position. This does not cause any difficulty when the quantum theory of atoms is made to accord with the special theory of relativity, because then gravitation is neglected, and the space-time is supposed to be flat whether there are atoms about in it or not. But if we try to make quantum theory accord with the general theory of relativity, then gravitation is not to be neglected, so that the curvature of space-time will depend on the whereabouts of the atoms. However, as we have just seen, the quantum theory makes it quite clear that we cannot always know where the atoms are. This is one root of the difficulty.

Finally we come to geography, in which I include history. The separation of history from geography rests upon the separation of time from space: when we amalgamate the two in space-time, we need one word to describe the combination of geography and history. For the sake of simplicity, I shall use the one word "geography" in this extended sense.

Geography, in this sense, includes everything that, as a matter of crude fact, distinguishes one part of space-time from another. One part is occupied by the sun, one by the earth; the intermediate regions contain light waves, but no matter (apart from a very little here and there). There is a certain degree of theoretical connection between different geographical facts; to establish this is the purpose of physical laws.

We are already in a position to calculate the large facts about the solar system backward and forward for vast periods of time. But in all such calculations we need a basis of crude fact. The facts are interconnected, but facts can only be inferred from other facts, not from general laws alone. Thus the facts of geography have a certain independent status in physics. No amount of physical laws will enable us to infer a physical fact unless we know other facts as data for our inference. And here when I speak of "facts" I am thinking of particular facts of geography, in the extended sense in which I am using the term.

In the theory of relativity, we are concerned with *structure,* not with the material of which the structure is composed. In geography, on the other hand, the material is relevant. If there is to be any difference between one place and another, there must either be differences between the material in one place and that in another, or places where there is material and places where there is none. The former of these alternatives seems the more satisfactory. We might try to say, There are electrons and protons and the other subatomic particles, and the rest is empty. But in the empty regions there are light waves, so that we cannot say that there is nothing there. According to quantum theory, we cannot even say exactly where things are, but only that one place is more likely than another to find an electron in. Some people maintain that light waves, and particles as well, are just disturbances in the ether, others are content to say that they are just disturbances; but in any case events are occurring wherever there are likely to be light waves or particles. That is all that we can say for the places where there is likely to be energy in one form or another, since energy has turned out to be a mathematical construction built out of events. We may say, therefore, that there are events everywhere in space-time, but they must be of a somewhat different kind according as we are dealing with a region where there is very likely to be an electron or proton, or with the sort of region we should ordinarily call empty. But as to the intrinsic nature of these events we can know nothing, except when they happen to be events in our own lives. Our own perceptions and feelings must be part of the crude material of events which physics arranges into a pattern—or rather, which physics finds to be arranged in a pattern. As regards events which do not form part of our own lives, physics tells us the pattern of them, but is quite unable to tell us what they are like in themselves. Nor does it seem possible that this should be discovered by any other method.

CHAPTER 13

The Abolition of "Force"

In the Newtonian system, bodies under the action of no forces move in straight lines with uniform velocity; when bodies do not move in this way, their change of motion is ascribed to a "force." Some forces seem intelligible to our imagination: those exerted by a rope or string, by bodies colliding, or by any kind of obvious pushing or pulling. As explained in an earlier chapter, our apparent imaginative understanding of these processes is quite fallacious; all that it really means is that past experience enables us to foresee more or less what is going to happen without the need of mathematical calculations. But the "forces" involved in gravitation and in the less-familiar forms of electrical action do not seem in this way "natural" to our imagination. It seems odd that the earth can float in the void: the natural thing to suppose is that it must fall. That is why it has to be supported on an elephant, and the elephant on a tortoise, according to some early speculators. The Newtonian theory, in addition to action at a distance, introduced two other imaginative novelties. The first was that gravitation is not always and essentially directed what we should call "downwards"—i.e., toward the center of the earth. The second was that a body going round and round in a circle with uniform velocity is not "moving uniformly" in the sense in which that phrase is applied to the motion of bodies under no forces, but is perpetually being turned out of the straight course toward the center of the circle, which requires a force pulling it in that direction. Hence Newton arrived at the view that the planets are attracted to the sun by a force, which is called gravitation.

This whole point of view, as we have seen, is superseded by relativity. There are no longer such things as "straight lines" in the old geometrical sense. There are "straightest lines," or geodesics, but these involve time as well as space. A light ray passing through the solar system does not describe the same orbit as a comet, from a geometrical point of view; nevertheless, each moves in a geodesic. The whole imaginative picture is changed. A poet might say that water runs downhill because it is attracted to the sea, but a physicist or an ordinary mortal would say that it moves as it does, at each point, because of the nature of the ground at that point, without regard to what lies ahead of it. Just as the sea does not cause the water to run toward it, so the sun does not cause the planets to move around it. The planets move around the sun because that is the easiest thing to do—in the technical sense of "least action." It is the easiest thing to do because of the nature of the region in which they are, not because of an influence emanating from the sun.

The supposed necessity of attributing gravitation to a "force" attracting the planets toward the sun has arisen from the determination to preserve Euclidean geometry at all costs. If we suppose that our space is Euclidean, when in fact it is not, we shall have to call in physics to rectify the errors of our geometry. We shall find bodies not moving in what we insist upon regarding as straight lines, and we shall demand a cause for this behavior. Eddington stated this matter with admirable lucidity and the following explanation is based on one given by him. Suppose that you assume the formula for interval which is used in the special theory of relativity—a formula which implies that your space is Euclidean. Since intervals can be compared by experimental methods, you will soon discover that your formula cannot be reconciled with the results of observation, and realize your mistake. If you insist on adhering to the Euclidean formula notwithstanding, then you will have to attribute the discrepancy between formula and observations to some influence which is present and which affects the behavior of test bodies. You will introduce an additional agency to which you can attribute the consequences of your mistake. The name given to any agency which causes deviation from uniform motion in a straight line is *force* according to the Newtonian definition of force. Hence the agency invoked through your insistence on the Euclidean formula for interval is described as a "field of force."

If people were to learn to conceive the world in the new way, without the old notion of "force," it would alter not only their physical imagination, but probably also their morals and politics. The latter effect would be quite illogical, but is not less probable on that account. In the Newtonian theory of the solar system, the sun seems like a monarch whose behests the planets have to obey. In the Einsteinian world there is more individualism and less government than in the Newtonian. There is also far less hustle: we have seen that laziness is the fundamental law of the Einsteinian universe. The word "dynamic" has come to mean, in newspaper language, "energetic and forceful";

The Abolition of "Force"

but if it meant "illustrating the principles of dynamics," it ought to be applied to people who sit under trees waiting for the fruit to drop into their mouths. I hope that journalists, in future, when they speak of a "dynamic personality," will mean a person who does what is least trouble at the moment, without thinking of remote consequences. If I can contribute to this result, I shall not have written in vain.

The abolition of "force" seems to be connected with the substitution of sight for touch as the source of physical ideas, as explained in Chapter 1. When an image in a looking glass moves, we do not think that something has pushed it. In places where there are two large mirrors opposite to each other, you may see innumerable reflections of the same object. Suppose that someone wearing a hat is standing between the mirrors, there may be twenty or thirty hats in the reflections. Suppose now somebody comes and knocks off the hat with a stick: all the other twenty or thirty hats will tumble down at the same moment. We think that a force is needed to knock off the "real" hat, but we think the remaining twenty or thirty tumble off, so to speak, of themselves, or out of a mere passion for imitation. Let us try to think out this matter a little more seriously.

Obviously something happens when an image in a looking glass moves. From the point of view of sight, the event seems just as real as if it were not in a mirror. But nothing has happened from the point of view of touch or hearing. When the "real" hat falls, it makes a noise; the twenty or thirty reflections fall without a sound. If it falls on your toe, you feel it, but we believe that the twenty or thirty people in the mirrors feel nothing, though hats fall on their toes too. But all this is equally true of the astronomical world. It makes no noise, because sound cannot travel across a vacuum. So far as we know, it causes no "feelings," because there is no one on the spot to "feel" it. The astronomical world, therefore, seems hardly more "real" or "solid" than the world in the looking glass, and has just as little need of "force" to make it move.

The reader may feel that I am indulging in idle sophistry. "After all," he may say, "the image in the mirror is the reflection of something solid, and the hat in the mirror falls off only because of the force applied to the real hat. The hat in the mirror cannot indulge in behavior of its own; it has to copy the real one. This shows how different the image is from the sun and the planets, because *they* are not obliged to be perpetually imitating a prototype. So you had better give up pretending that an image is just as real as one of the heavenly bodies."

There is, of course, some truth in this; the point is to discover exactly *what* truth. In the first place, images are not "imaginary." When you see an image, certain perfectly real light

The Abolition of "Force"

waves reach your eye, and if you hang a cloth over the mirror, these light waves cease to exist. There is, however, a purely optical difference between an "image" and a "real" thing. The optical difference is bound up with this question of imitation. When you hang a cloth over the mirror, it makes no difference to the "real" object, but when you move the "real" object away, the image vanishes also. This makes us say that the light rays which make the image are only reflected at the surface of the mirror, and do not really come from a point behind it, but from the "real" object. We have here an example of a general principle of great importance. Most of the events in the world are not isolated occurrences, but members of groups of more or less similar events, which are such that each group is connected in an assignable manner with a certain small region of space-time. This is the case with the light rays which make us see both the object and its reflection in the mirror: they all emanate from the object as a center. If you put an opaque globe around the object at a certain distance, the object and its reflection are invisible at any point outside the globe. We have seen that gravitation, although no longer regarded as an action at a distance, is still connected with a center: there is, so to speak, a hill symmetrically arranged about its summit, and the summit is the place where we conceive the body to be which is connected with the gravitational field we are considering. For simplicity, common sense lumps together all the events which form one group in the above sense. When two people see the same object, two different events occur, but they are events belonging to one group and connected with the same center. Just the same applies when two people (as we say) hear the same noise. And so the reflection in a mirror is less "real" than the object reflected, even from an optical point of view, because light rays do not spread in *all* directions from the place where the image seems to be, but only in directions in front of the mirror, and only so long as the object reflected remains in position. This illustrates the usefulness of grouping connected events about a center in the way we have been considering.

When we examine the changes in such a group of objects, we find that they are of two kinds: there are those which affect only some member of the group, and those which make connected alterations in all the members of the group. If you put a candle in front of a mirror, and then hang black cloth over the mirror, you alter only the reflection of the candle as seen from various places. If you shut your eyes, you alter its appearance to you, but not its appearance elsewhere. If you put a red globe around it at a distance of a foot, you alter its appearance at any distance greater than a foot, but not at any distance less than a foot. In all these cases, you do not regard the candle itself as having changed; in fact, in all of them you find that there are groups of changes connected with a different center or with a number of different centers. When you shut your eyes, for instance, your eyes, not the candle, look different to any other observer: the center of the changes that occur is in your eyes. But when you blow out the candle, its appearance *everywhere* is changed; in this case you say that the change has happened to the candle. The changes that happen to an object are those that affect the whole group of events which center about the object. All this is only an interpretation of common sense and an attempt to explain what we mean by saying that the image of the candle in the mirror is less "real" than the candle. There is no connected group of events situated all around the place where the image seems to be, and changes in the image center about the candle, not about a point about the mirror. This gives a perfectly verifiable meaning to the statement that the image is "only" a reflection. And at the same time it enables us to regard the heavenly bodies, although we can only see and not touch them, as more "real" than an image in a looking glass.

We can now begin to interpret the common-sense notion of one body having an "effect" upon another, which we must do if we are really to understand what is meant by the abolition of "force." Suppose you come into a dark room and switch on the electric light: the appearance of everything in the room is changed. Since everything in the room is visible because it reflects the electric light, this case is really analogous to that of the image in the mirror; the electric light is the center from which all the changes emanate. In this case, the "effect" is explained by what we have already said. The more important case is when the effect is a movement. Suppose you let loose a tiger in the middle of a bank holiday crowd: they would all move, and the tiger would be the center of their various movements. A person who could see the people but not the tiger would infer that there was something repulsive at that point. We say in this case that the tiger has an effect upon the people, and we might describe the tiger's action upon them as of the nature of a repulsive force. We know, however, that they fly because of something which happens to *them*, not merely because the tiger is where it is. They fly because they can see and hear it, that is to say, because certain waves reach their eyes and ears. If these waves could be made to reach them without there being any tiger, they would fly just as fast, because the neighborhood would seem to them just as unpleasant.

Let us now apply similar considerations to the sun's gravitation. The "force" exerted by the sun only differs from that exerted by the tiger in being attractive instead of repulsive. Instead of acting through waves of light or sound, the sun acquires its apparent power through the fact that there are modifications of space-time all around the sun. Like the noise of the tiger, they are more intense near their source; as we travel away, they grow less and less. To say that the sun "causes" these modifications of space-time is to add nothing to our knowledge. What we know is that the modifications proceed according to a certain rule, and that they are grouped symmetrically about the sun as center. The language of cause and effect adds only a number of quite irrelevant imaginings, connected with will, muscular tension, and such matters. What we can more or less ascertain is merely the formula according to which space-time is modified by the presence of gravitating matter. More correctly: we can ascertain what kind of space-time *is* the presence of gravitating matter. When space-time is not accurately Euclidean in a certain region, but has a non-Euclidean character which grows more and more marked as we approach a certain center, and when, further, the departure from Euclid obeys a certain law, we describe this state of affairs briefly by saying that there is gravitating matter at the center. But this is only a compendious account of what we know. What we know is about the places where the gravitating matter is *not*, not about the place where it is. The language of cause and effect (of which "force" is a particular case) is thus merely a convenient shorthand for certain purposes; it does not represent anything that is genuinely to be found in the physical world.

And how about matter? Is matter also no more than a convenient shorthand? This question, however, being a large one, demands a separate chapter.

CHAPTER 14

What Is Matter?

The question "What is matter?" is of the kind that is asked by metaphysicians, and answered in vast books of incredible obscurity. But I am not asking the question as a metaphysician: I am asking it as a person who wants to find out what is the moral of modern physics, and more especially of the theory of relativity. It is obvious from what we have learned of that theory that matter cannot be conceived quite as it used to be. I think we can now say more or less what the new conception must be.

There were two traditional conceptions of matter, both of which have had advocates ever since scientific speculation began. There were the atomists, who thought that matter consisted of tiny lumps which could never be divided; these were supposed to hit each other and then bounce off in various ways. After Newton, they were no longer supposed actually to come into contact with each other, but to attract and repel each other, and move in orbits around each other. Then there were those who thought that there is matter of some kind everywhere, and that a true vacuum is impossible. Descartes held this view and attributed the motions of the planets to vortices in the ether. The Newtonian theory of gravitation caused the view that there is matter everywhere to fall into discredit, the more so as light was thought by Newton and his successors to be due to actual particles traveling from the source of the light. But when this view of light was disproved, and it was shown that light consisted of waves, the ether was revived so that there should be something to undulate. The ether became still more respectable when it was found to play the same part in electromagnetic phenomena as in the propagation of light. It was even hoped that atoms might turn out to be a mode of motion of the ether. At this stage, the atomic view of matter was, on the whole, getting the worst of it.

Leaving relativity aside for the moment, modern physics has provided proof of the atomic structure of ordinary matter, while not disproving the arguments in favor of the ether, to which no such structure is attributed. The result was a sort of compromise between the two views, the one applying to what was called "gross" matter, the other to the ether. There can be no doubt about electrons and protons, though, as we shall see shortly, they need not be conceived as atoms were conceived traditionally. The truth is, I think, that relativity demands the abandonment of the old conception of "matter," which is infected by the metaphysics associated with "substance," and represents a point of view not really necessary in dealing with phenomena. This is what we must now investigate.

In the old view, a piece of matter was something which survived all through time, while never being at more than one place at a given time. This way of looking at things is obviously connected with the complete separation of space and time in which people formerly believed. When we substitute space-time for space and time, we shall naturally expect to derive the physical world from constituents which are as limited in time as in space. Such constituents are what we call "events." An event does not persist and move, like the traditional piece of matter; it merely exists for its little moment and then ceases. A piece of matter will thus be resolved into a series of events. Just as, in the old view, an extended body was composed of a number of particles, so, now, each particle, being extended in time, must be regarded as composed of what we may call "event-particles." The whole series of these events makes up the whole history of the particle, and the particle is regarded as *being* its history, not some metaphysical entity to which the events happen. This view is rendered necessary by the fact that relativity compels us to place time and space more on a level than they were in the older physics.

This abstract requirement must be brought into relation with the known facts of the physical world. Now what are the known facts? Let us take it as conceded that light consists of waves traveling with the received velocity. We then know a great deal about what goes on in the parts of space-time where there is no matter; we know, that is to say, that there are periodic occurrences (light waves) obeying certain laws. These light waves start from atoms, and the modern theory of the structure of the atom enables us to know a great deal about the circumstances under which they start and the reasons which determine their wavelengths. We can find out not only how one light wave travels, but how its source moves relatively to ourselves. But when I say this I am assuming that we can recognize a source of light as the same at two slightly different times. This is, however, the very thing which had to be investigated.

We saw, in the preceding chapter, how a

group of connected events can be formed, all related to each other by a law and all ranged about a center in space-time. Such a group of events will be the arrival, at various places, of the light waves emitted by a brief flash of light. We do not need to suppose that anything particular is happening at the center; certainly we do not need to suppose that we know *what* is happening there. What we know is that, as a matter of geometry, the group of events in question are ranged about a center, like widening ripples on a pool when a fly has touched it. We can hypothetically invent an occurrence which is to have happened at the center, and set forth laws by which the consequent disturbance is transmitted. This hypothetical occurrence will then appear to common sense as the "cause" of the disturbance. It will also count as one event in the biography of the particle of matter which is supposed to occupy the center of the disturbance.

Now we find not only that one light wave travels outward from a center according to a certain law, but also that, in general, it is followed by other closely similar light waves. The sun, for example, does not change its appearance suddenly; even if a cloud passes across it during a high wind, the transition is gradual, though swift. In this way a group of occurrences connected with a center at one point of space-time is brought into relation with other very similar groups whose centers are at neighboring points of space-time. For each of these other groups common sense invents similar hypothetical occurrences to occupy their centers, and says that all these hypothetical occurrences are part of one history—that is to say, it invents a hypothetical "particle" to which the hypothetical occurrences are to have occurred. It is only by this double use of hypothesis, perfectly unnecessary in each case, that we arrive at anything that can be called "matter" in the old sense of the word.

If we are to avoid unnecessary hypotheses, we shall say that an atom at a given moment *is* the various disturbances in the surrounding medium which, in ordinary language, would be said to be "caused" by it. But we shall not take these disturbances at what is, for us, the moment in question, since that would make them depend upon the observer; we shall instead travel outward from the atom with the velocity of light and take the disturbance we find in each place as we reach it. The closely similar set of disturbances, with very nearly the same center, which is found existing slightly earlier or slightly later, will be defined as *being* the atom at a slightly earlier or slightly later moment. In this way, we preserve all the laws of physics without having recourse to unnecessary hypotheses or inferred entities, and we remain in harmony with the general principle of economy which has enabled the theory of relativity to clear away so much useless lumber.

Common sense imagines that when it sees a table it sees a table. This is a gross delusion. When common sense sees a table, certain light waves reach its eyes, and these are of a sort which, in its previous experience, has been associated with certain sensations of touch, as well as with other people's testimony that they also saw the table. But none of this ever brought us to the table itself. The light waves caused occurrences in our eyes, and these caused occurrences in the optic nerve, and these in turn caused occurrences in the brain. Any one of these, happening without the usual preliminaries, would have caused us to have the sensations we call "seeing the table," even if there had been no table. (Of course, if matter in general is to be interpreted as a group of occurrences, this must apply also to the eye, the optic nerve, and the brain.) As to the sense of touch when we press the table with our fingers, that is an electric disturbance on the electrons and protons of our fingertips, produced, according to modern physics, by the proximity of the electrons and protons in the table. If the same disturbance in our fingertips arose in any other way, we should have the sensations, in spite of there being no table. The testimony of others is obviously a secondhand affair. A witness in a law court, if asked whether she or he had seen some occurrence, would not be allowed to reply that she or he believed so because of the testimony of others to that effect. In any case, testimony consists of sound waves and demands psychological as well as physical interpretation; its connection with the object is therefore very indirect. For all these reasons, when we say that a person "sees a table," we use a highly abbreviated form of expression, concealing complicated and difficult inferences, the validity of which may well be open to question.

But we are in danger of becoming entangled in psychological questions, which we must avoid if we can. Let us therefore return to the purely physical point of view.

What I wish to suggest may be put as follows: everything that occurs elsewhere, owing to the existence of an atom, can be explored experimentally, at least in theory, unless it occurs in certain concealed ways. An atom is known by its "effects." But the word "effects" belongs to a view of causation which will not fit modern physics, and in particular will not fit relativity. All that we have a right to say is that certain groups of occurrences happen together, that is to say, in neighboring parts of space-time. A given observer will regard one member of the group as earlier than the other, but another observer may judge the time-order differently. And even when the time-order is the same for all observers, all that we really have is a connection between the two events, which works equally backward and forward. It is not true that the past determines the future in some sense other than that in which the future determines the past: the apparent difference is only due to our ignorance, because we know less about the future than about the past. This is a mere accident: there might be beings who would remember the future and have to infer the past. The feelings of such beings in these matters would be the exact opposite of our own, but no more fallacious.

It seems fairly clear that all the facts and laws of physics can be interpreted without assuming that "matter" is anything more than groups of events, each event being of the sort which we should naturally regard as "caused" by the matter in question. This does not involve any change in the symbols or formulas of physics: it is merely a question of interpretation of the symbols.

This latitude in interpretation is a characteristic of mathematical physics. What we know is certain very abstract logical relations, which we express in mathematical formulas; we know also that, at certain points, we arrive at results which are capable of being tested experimentally. Take, for example, the astronomical observations by which the predictions of relativity theory as to the behavior of light were confirmed. The formulas which were to be verified were concerned with the course of light in interplanetary space. Although the part of these formulas which gives the observed result must always be interpreted in the same way, the other part of them may be capable of a great variety of interpretations. The formulas giving the motions of the planets are almost exactly the same in Einstein's theory as in Newton's, but the meaning of the formulas is quite different. It may be said generally that, in the mathematical treatment of nature, we can be far more certain that our formulas are approximately correct than we can be as to the correctness of this or that interpretation of them. And so in the case with which this chapter is concerned; the question as to the nature of an electron or a proton is by no means answered when we know all that mathematical physics has to say as to the laws of its motion and the laws of its interaction with the environment. A definite and conclusive answer to our question is not possible, just because a variety of answers are compatible with the truth of mathematical physics. Nevertheless, some answers are preferable to others, because some have a greater probability in their favor. We have been seeking, in this chapter, to define matter so that there *must* be such a thing, if the formulas of physics are true. If we had made our definition such as to secure that a particle of matter should be what one thinks of as substantial, a hard, definite lump, we should

What Is Matter?

not have been *sure* that any such thing exists. That is why our definition, though it may seem complicated, is preferable from the point of view of logical economy and scientific caution.

CHAPTER 15

Philosophical Consequences

The philosophical consequences of relativity are neither so great nor so startling as is sometimes thought. It throws very little light on time-honored controversies, such as that between realism and idealism. Some people think that it supports Kant's view that space and time are "subjective" and are "forms of intuition." I think such people have been misled by the way in which writers on relativity speak of "the observer." It is natural to suppose that the observer is a human being, or at least a mind, but it is just as likely to be a photographic plate or a clock. That is to say, the odd results as to the difference between one "point of view" and another are concerned with "point of view" in a sense applicable to physical instruments just as much as to people with perceptions. The "subjectivity" concerned in the theory of relativity is a *physical* subjectivity, which would exist equally if there were no such things as minds or senses in the world.

Moreover, it is a strictly limited subjectivity. The theory does not say that *everything* is relative; on the contrary, it gives a technique for distinguishing what is relative from what belongs to a physical occurrence in its own right. If we are going to say that the theory supports Kant about space and time, we shall have to say that it refutes him about space-time. In my view, neither statement is correct. I see no reason why, on such issues, philosophers should not all stick to the views they previously held. There were no conclusive arguments on either side before, and there are none now; to hold either view shows a dogmatic rather than a scientific temper.

Nevertheless, when the ideas involved in the theory of relativity have become familiar, as they will do when they are taught in schools, certain changes in our habits of thought are likely to result, and to have great importance in the long run.

One thing which emerges is that physics tells us much less about the physical world than we thought it did. Almost all the "great principles" of traditional physics turn out to be like the "great law" that there are always three feet to a yard; others turn out to be downright false. The conservation of mass may serve to illustrate both these misfortunes to which a "law" is liable. Mass used to be defined as "quantity of matter," and as far as experiment showed, it was never increased or diminished. But with the greater accuracy of modern measurements, curious things were found to happen. In the first place, the mass as measured was found to increase with the velocity; this kind of mass was found to be really the same thing as energy. This kind of mass is not constant for a given body. The law itself, however, is to be regarded as a truism, of the nature of the "law" that there are three feet to a yard; it results from our methods of measurement and does not express a genuine property of matter. The other kind of mass, which we may call "proper mass," is that which is found to be the mass by an observer moving with the body. This is the ordinary terrestrial case where the body we are weighing is not flying through the air. The "proper mass" of a body is very nearly constant, but not quite. One would suppose that if you have four one-pound weights, and you put them all together into the scales, they will together weigh four pounds. This is a fond delusion: they weigh rather less, though not enough less to be discovered by even the most careful measurements. In the case of four hydrogen atoms, however, when they are put together to make one helium atom, the defect is noticeable; the helium atom weights measurably less than four separate hydrogen atoms.

Broadly speaking, traditional physics has collapsed into two portions, truisms and geography.

The world which the theory of relativity presents to our imagination is not so much a world of "things" in "motion" as a world of *events*. It is true that there are still particles which seem to persist, but these (as we saw in the preceding chapter) are really to be conceived as strings of connected events, like the successive notes of a song. It is *events* that are the stuff of relativity physics. Between two events which are not too remote from each other there is, in the general theory as in the special theory, a measurable relation called "interval," which appears to be the physical reality of which lapse of time and distance in space are two more or less confused representations. Between two distant events, there is not any one definite interval. But there is one way of moving from one event to another which makes the sum of all the little intervals along the route greater than by any other route. This route is called a "geodesic," and it is the route which a body will choose if left to itself.

The whole of relativity physics is a much

more step-by-step matter than the physics and geometry of former days. Euclidean straight lines have to be replaced by light rays, which do not quite come up to the Euclidean standard of straightness when they pass near the sun or any other very heavy body. The sum of the angles of a triangle is still thought to be two right angles in very small regions of empty space, but not in any extended region. Nowhere can we find a place where Euclidean geometry is exactly true. Propositions which used to be proved by reasoning have now become either conventions, or merely approximate truths verified by observation.

It is a curious fact—of which relativity is not the only illustration—that, as reasoning improves, its claims to the power of proving facts grow less and less. Logic used to be thought to teach us how to draw inferences; now, it teaches us rather how not to draw inferences. Animals and children are terribly prone to inference: a horse is surprised beyond measure if you take an unusual turning. When men began to reason, they tried to justify the inferences that they had drawn unthinkingly in earlier days. A great deal of bad philosophy and bad science resulted from this propensity. "Great principles," such as the "uniformity of nature," the "law of universal causation," and so on, are attempts to bolster up our belief that what has often happened before will happen again, which is no better founded than the horse's belief that you will take the turning you usually take. It is not altogether easy to see what is to replace these pseudoprinciples in the practice of science, but perhaps the theory of relativity gives us a glimpse of the kind of thing we may expect. Causation, in the old sense, no longer has a place in theoretical physics. There is, of course, something else which takes its place, but the substitute appears to have a better empirical foundation than the old principle which it has superseded.

The collapse of the notion of one all-embracing time, in which all events throughout the universe can be dated, must in the long run affect our views as to cause and effect, evolution, and many other matters. For instance, the question whether, on the whole, there is progress in the universe may depend upon our choice of a measure of time. If we choose one out of a number of equally good clocks, we may find that the universe is progressing as fast as the most optimistic American thinks it is; if we choose another equally good clock, we may find that the universe is going from bad to worse as fast as the most melancholy Slav could imagine. Thus optimism and pessimism are neither true nor false, but depend upon the choice of clocks.

The effect of this upon a certain type of emotion is devastating. The poet speaks of

> One far-off divine event
> To which the whole creation moves.

But if the event is sufficiently far off and the creation moves sufficiently quickly, some parts will judge that the event has already happened, while others will judge that it is still in the future. This spoils the poetry. The second line ought to be:

> To which some parts of the creation move, while others move away from it.

But this won't do. I suggest that an emotion which can be destroyed by a little mathematics is neither very genuine nor very valuable. But this line of argument would lead to a criticism of the Victorian age, which lies outside my theme.

What we know about the physical world, I repeat, is much more abstract than was formerly supposed. Between bodies there are occurrences, such as light waves; of the *laws* of these occurrences we know something—just so much as can be expressed in mathematical formulas—but of their *nature* we know nothing. Of the bodies themselves, as we saw in the preceding chapter, we know so little that we cannot even be sure that they are anything: they *may* be merely groups of events in other places, those events which we should naturally regard as their effects. We naturally interpret the world pictorially—that is to say, we imagine that what goes on is more or less like what we see. But in fact this likeness can only extend to certain formal logical properties expressing structure, so that all we can know is certain general characteristics of its changes. Perhaps an illustration may make the matter clear. Between a piece of orchestral music as played, and the same piece of music as printed in the score, there is a certain resemblance, which may be described as a resemblance of structure. The resemblance is of such a sort that, when you know the rules, you can infer the music from the score or the score from the music. But suppose you had been stone-deaf from birth, but had lived among musical people. You could understand, if you had learned to speak and to do lip-reading, that the musical scores represented something quite different from themselves in intrinsic quality, though similar in structure.* The value of music would be completely unimaginable to you, but you could infer all its mathematical characteristics, since they are the same as those of the score. Now, our knowledge of nature is something like this. We can read the scores and infer just so much as our stone-deaf person could have

---

*For the definition of "structure" see the present author's *Introduction to Mathematical Philosophy.*

inferred about music. But we have not the advantages which that person derived from association with musical people. We cannot know whether the music represented by the scores is beautiful or hideous; perhaps, in the last analysis, we cannot be quite sure that the scores represent anything but themselves. But this is a doubt which the physicist, in a professional capacity, cannot entertain.

Assuming the utmost that can be claimed for physics, it does not tell us what it is that changes, or what are its various states; it only tells us such things as that changes follow each other periodically, or spread with a certain speed. Even now we are probably not at the end of the process of stripping away what is merely imagination in order to reach the core of true scientific knowledge. The theory of relativity has accomplished a very great deal in this respect, and in doing so has taken us nearer and nearer to bare structure, which is the mathematicians' goal—not because it is the only thing which interests them as human beings, but because it is the only thing that they can express in mathematical formulas. But far as we have traveled in the direction of abstraction, it may be that we shall have to travel farther still.

In the preceding chapter, I suggested what may be called a minimum definition of matter— that is to say, one in which matter has, so to speak, as little "substance" as is compatible

Philosophical Consequences

with the truth of physics. In adopting a definition of this kind, we are playing for safety: our tenuous matter will exist, even if something more beefy also exists. We tried to make our definition of matter, like Isabella's gruel in Jane Austen, "thin, but not too thin." We shall, however, fall into error if we assert positively that matter is nothing more than this. Leibniz thought that a piece of matter is really a colony of souls. There is nothing to show that he was wrong, though there is also nothing to show that he was right: we know no more about it either way than we do about the flora and fauna of Mars.

To the nonmathematical mind, the abstract character of our physical knowledge may seem unsatisfactory. From an artistic or imaginative point of view, it is perhaps regrettable, but from a practical point of view it is of no consequence. Abstraction, difficult as it is, is the source of practical power. A financier, whose dealings with the world are more abstract than those of any other "practical" person, is also more powerful than any other practical person. Financiers can deal in wheat or cotton without needing ever to have seen either: all they need to know is whether the price will go up or down. This is abstract mathematical knowledge, at least as compared to the knowledge of the agriculturist. Similarly, the physicist, who knows nothing of matter except certain laws of its movements, nevertheless knows enough to be able to manipulate it. After working through whole strings of equations, in which the symbols stand for things whose intrinsic nature can never be known to us, the physicist arrives at last at a result which can be interpreted in terms of our own perceptions, and utilized to bring about desired effects in our own lives. What we know about matter, abstract and schematic as it is, is enough, in principle, to tell us the rules according to which it produces perceptions and feelings in ourselves, and it is upon these rules that the *practical* uses of physics depend.

The final conclusion is that we know very little, and yet it is astonishing that we know so much, and still more astonishing that so little knowledge can give us so much power.

Nature and Man from MENTOR

MENTOR Books of Special Interest

MENTOR Books of Plays

(0451)

☐ **EIGHT GREAT TRAGEDIES edited by Sylvan Barnet, Morton Berman and William Burto.** The great dramatic literature of the ages. Eight memorable tragedies by Aeschylus, Euripides, Sophocles, Shakespeare, Ibsen, Strindberg, Yeats, and O'Neill. With essays on tragedy by Aristotle, Emerson and others. (62258—$3.50)

☐ **EIGHT GREAT TRAGEDIES edited by Sylvan Barnet, Morton Berman and William Burto.** Complete texts of eight masterpieces of comic drama by Aristophanes, Machiavelli, Shakespeare, Molière, John Gay, Wilde, Chekhov, and Shaw. Includes essays on comedy by four distinguished critics and scholars. (123649—$4.95)

☐ **THE GENIUS OF THE EARLY ENGLISH THEATRE edited by Sylvan Barnet, Morton Berman and William Burto.** Complete plays including three anonymous plays —"Abraham and Isaac," "The Second Shepherd's Play," and "Everyman." and Marlowe's "Doctor Faustus," Shakespeare's "Macbeth." Jonson's "Volpone," and Milton's "Samson Agonistes," with critical essays. (622219—$3.95)

☐ **THE MENTOR BOOK OF SHORT PLAYS edited by Richard Goldstone and Abraham H. Lass.** Introduction and "On Reading Plays" by the editors. Includes: *A Trip to Czardis.* Granberry: *To Bobolink for Her Spirit.* Inge; *Riders to the Sea.* Synge; *Visit to a Small Planet.* Vidal; *The Mother,* Chayefsky: *Thunder on Sycamore Street.* Rose; *The Rising of the Moon,* Gregory; *The Happy Journey.* Wilde; *Lord Byron's Love Letter,* Williams; *The Marriage Proposal. Chekhov: The Romancers,* Rostand; *The Browning Version.* Rattigan. (622154—$3.95)